T0350086

Graduate Texts in Mathematics 41

Springer

New York
Berlin
Heidelberg
Barcelona
Hong Kong
London
Milan
Paris
Singapore
Tokyo

Graduate Texts in Mathematics

(continued after index)

Tom M. Apostol

Modular Functions and Dirichlet Series in Number Theory

Second Edition

With 25 Illustrations

Springer

Tom M. Apostol
Department of Mathematics
California Institute of Technology
Pasadena, CA 91125
USA

Mathematics Subject Classification (2000): 11-01, 11FXX

Library of Congress Cataloging-in-Publication Data
Apostol, Tom M.
    Modular functions and Dirichlet series in number theory/Tom M. Apostol.—2nd ed.
      p. cm.—(Graduate texts in mathematics; 41)
    Includes bibliographical references.
    ISBN 0-387-97127-0 (alk. paper)
      1. Number theory. 2. Functions, Elliptic. 3. Functions, Modular. 4. Series,
Dirichlet. I. Title. II. Series.
QA241.A62   1990
512′.7—dc20   89-21760

Printed on acid-free paper.

Typeset by Asco Trade Typesetting Ltd., Hong Kong.
Printed and bound by Berryville Graphics, Inc., Berryville, VA.
Printed in the United States of America.

9 8 7 6 5 4 3

ISBN 0-387-97127-0
ISBN 3-540-97127-0   SPIN 10841555

Springer-Verlag New York Berlin Heidelberg
A member of BertelsmannSpringer Science+Business Media GmbH

Preface

This is the second volume of a 2-volume textbook* which evolved from a course (Mathematics 160) offered at the California Institute of Technology during the last 25 years.

The second volume presupposes a background in number theory comparable to that provided in the first volume, together with a knowledge of the basic concepts of complex analysis.

Most of the present volume is devoted to elliptic functions and modular functions with some of their number-theoretic applications. Among the major topics treated are Rademacher's convergent series for the partition function, Lehner's congruences for the Fourier coefficients of the modular function $j(\tau)$, and Hecke's theory of entire forms with multiplicative Fourier coefficients. The last chapter gives an account of Bohr's theory of equivalence of general Dirichlet series.

Both volumes of this work emphasize classical aspects of a subject which in recent years has undergone a great deal of modern development. It is hoped that these volumes will help the nonspecialist become acquainted with an important and fascinating part of mathematics and, at the same time, will provide some of the background that belongs to the repertory of every specialist in the field.

This volume, like the first, is dedicated to the students who have taken this course and have gone on to make notable contributions to number theory and other parts of mathematics.

T. M. A.
January, 1976

* The first volume is in the Springer-Verlag series Undergraduate Texts in Mathematics under the title *Introduction to Analytic Number Theory*.

Preface to the Second Edition

The major change is an alternate treatment of the transformation formula for the Dedekind eta function, which appears in a five-page supplement to Chapter 3, inserted at the end of the book (just before the Bibliography). Otherwise, the second edition is almost identical to the first. Misprints have been repaired, there are minor changes in the Exercises, and the Bibliography has been updated.

T. M. A.
July, 1989

Contents

Chapter 3

The Dedekind eta function

Chapter 4

Congruences for the coefficients of the modular function j

Chapter 5

Rademacher's series for the partition function

Elliptic functions 1

1.1 Introduction

Additive number theory is concerned with expressing an integer n as a sum of integers from some given set S. For example, S might consist of primes, squares, cubes, or other special numbers. We ask whether or not a given number can be expressed as a sum of elements of S and, if so, in how many ways this can be done.

Let $f(n)$ denote the number of ways n can be written as a sum of elements of S. We ask for various properties of $f(n)$, such as its asymptotic behavior for large n. In a later chapter we will determine the asymptotic value of the partition function $p(n)$ which counts the number of ways n can be written as a sum of positive integers $\leq n$.

The partition function $p(n)$ and other functions of additive number theory are intimately related to a class of functions in complex analysis called *elliptic modular functions*. They play a role in additive number theory analogous to that played by Dirichlet series in multiplicative number theory. The first three chapters of this volume provide an introduction to the theory of elliptic modular functions. Applications to the partition function are given in Chapter 5.

We begin with a study of doubly periodic functions.

1.2 Doubly periodic functions

A function f of a complex variable is called *periodic* with period ω if

$$f(z + \omega) = f(z)$$

whenever z and $z + \omega$ are in the domain of f. If ω is a period, so is $n\omega$ for every integer n. If ω_1 and ω_2 are periods, so is $m\omega_1 + n\omega_2$ for every choice of integers m and n.

Definition. A function f is called *doubly periodic* if it has two periods ω_1 and ω_2 whose ratio ω_2/ω_1 is not real.

We require that the ratio be nonreal to avoid degenerate cases. For example, if ω_1 and ω_2 are periods whose ratio is real and rational it is easy to show that each of ω_1 and ω_2 is an integer multiple of the same period. In fact, if $\omega_2/\omega_1 = a/b$, where a and b are relatively prime integers, then there exist integers m and n such that $mb + na = 1$. Let $\omega = m\omega_1 + n\omega_2$. Then ω is a period and we have

$$\omega = \omega_1\left(m + n\frac{\omega_2}{\omega_1}\right) = \omega_1\left(m + n\frac{a}{b}\right) = \frac{\omega_1}{b}(mb + na) = \frac{\omega_1}{b},$$

so $\omega_1 = b\omega$ and $\omega_2 = a\omega$. Thus both ω_1 and ω_2 are integer multiples of ω.

If the ratio ω_2/ω_1 is real and irrational it can be shown that f has arbitrarily small periods (see Theorem 7.12). A function with arbitrarily small periods is constant on every open connected set on which it is analytic. In fact, at each point of analyticity of f we have

$$f'(z) = \lim_{z_n \to 0} \frac{f(z + z_n) - f(z)}{z_n},$$

where $\{z_n\}$ is any sequence of nonzero complex numbers tending to 0. If f has arbitrarily small periods we can choose $\{z_n\}$ to be a sequence of periods tending to 0. Then $f(z + z_n) = f(z)$ and hence $f'(z) = 0$. In other words, $f'(z) = 0$ at each point of analyticity of f, hence f must be constant on every open connected set in which f is analytic.

1.3 Fundamental pairs of periods

Definition. Let f have periods ω_1, ω_2 whose ratio ω_2/ω_1 is not real. The pair (ω_1, ω_2) is called a *fundamental pair* if every period of f is of the form $m\omega_1 + n\omega_2$, where m and n are integers.

Every fundamental pair of periods ω_1, ω_2 determines a network of parallelograms which form a tiling of the plane. These are called *period parallelograms*. An example is shown in Figure 1.1a. The vertices are the periods $\omega = m\omega_1 + n\omega_2$. It is customary to consider two intersecting edges and their point of intersection as the only boundary points belonging to the period parallelogram, as shown in Figure 1.1b.

Notation. If ω_1 and ω_2 are two complex numbers whose ratio is not real we denote by $\Omega(\omega_1, \omega_2)$, or simply by Ω, the set of all linear combinations $m\omega_1 + n\omega_2$, where m and n are arbitrary integers. This is called the *lattice* generated by ω_1 and ω_2.

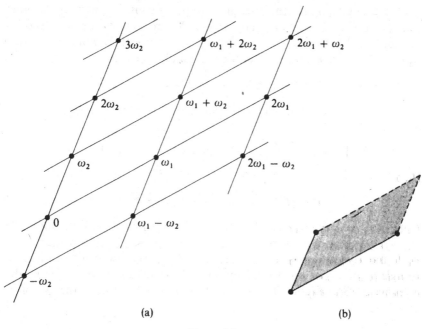

Figure 1.1

Theorem 1.1. *If* (ω_1, ω_2) *is a fundamental pair of periods, then the triangle with vertices* $0, \omega_1, \omega_2$ *contains no further periods in its interior or on its boundary. Conversely, any pair of periods with this property is fundamental.*

PROOF. Consider the parallelogram with vertices $0, \omega_1, \omega_1 + \omega_2$, and ω_2, shown in Figure 1.2a. The points inside or on the boundary of this parallelogram have the form

$$z = \alpha\omega_1 + \beta\omega_2,$$

where $0 \le \alpha \le 1$ and $0 \le \beta \le 1$. Among these points the only periods are 0, ω_1, ω_2, and $\omega_1 + \omega_2$, so the triangle with vertices $0, \omega_1, \omega_2$ contains no periods other than the vertices.

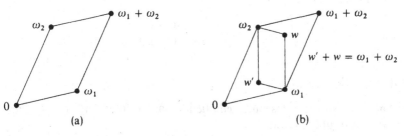

Figure 1.2

3

Conversely, suppose the triangle $0, \omega_1, \omega_2$ contains no periods other than the vertices, and let ω be any period. We are to show that $\omega = m\omega_1 + n\omega_2$ for some integers m and n. Since ω_2/ω_1 is nonreal the numbers ω_1 and ω_2 are linearly independent over the real numbers, hence

$$\omega = t_1\omega_1 + t_2\omega_2$$

where t_1 and t_2 are real. Now let $[t]$ denote the greatest integer $\leq t$ and write

$$t_1 = [t_1] + r_1, t_2 = [t_2] + r_2, \text{ where } 0 \leq r_1 < 1 \text{ and } 0 \leq r_2 < 1.$$

Then

$$\omega - [t_1]\omega_1 - [t_2]\omega_2 = r_1\omega_1 + r_2\omega_2.$$

If one of r_1 or r_2 is nonzero, then $r_1\omega_1 + r_2\omega_2$ will be a period lying inside the parallelogram with vertices $0, \omega_1, \omega_2, \omega_1 + \omega_2$. But if a period w lies inside this parallelogram then either w or $\omega_1 + \omega_2 - w$ will lie inside the triangle $0, \omega_1, \omega_2$ or on the diagonal joining ω_1 and ω_2, contradicting the hypothesis. (See Figure 1.2b.) Therefore $r_1 = r_2 = 0$ and the proof is complete. $\qquad\square$

Definition. Two pairs of complex numbers (ω_1, ω_2) and (ω_1', ω_2'), each with nonreal ratio, are called *equivalent* if they generate the same lattice of periods; that is, if $\Omega(\omega_1, \omega_2) = \Omega(\omega_1', \omega_2')$.

The next theorem, whose proof is left as an exercise for the reader, describes a fundamental relation between equivalent pairs of periods.

Theorem 1.2. *Two pairs (ω_1, ω_2) and (ω_1', ω_2') are equivalent if, and only if, there is a 2×2 matrix $\begin{pmatrix} a & b \\ c & d \end{pmatrix}$ with integer entries and determinant $ad - bc = \pm 1$, such that*

$$\begin{pmatrix} \omega_2' \\ \omega_1' \end{pmatrix} = \begin{pmatrix} a & b \\ c & d \end{pmatrix}\begin{pmatrix} \omega_2 \\ \omega_1 \end{pmatrix},$$

or, in other words,

$$\omega_2' = a\omega_2 + b\omega_1,$$
$$\omega_1' = c\omega_2 + d\omega_1.$$

1.4 Elliptic functions

Definition. A function f is called *elliptic* if it has the following two properties:
(a) f is doubly periodic.
(b) f is meromorphic (its only singularities in the finite plane are poles).

Constant functions are trivial examples of elliptic functions. Later we shall give examples of nonconstant elliptic functions, but first we derive some fundamental properties common to all elliptic functions.

Theorem 1.3. *A nonconstant elliptic function has a fundamental pair of periods.*

PROOF. If f is elliptic the set of points where f is analytic is an open connected set. Also, f has two periods with nonreal ratio. Among all the nonzero periods of f there is at least one whose distance from the origin is minimal (otherwise f would have arbitrarily small nonzero periods and hence would be constant). Let ω be one of the nonzero periods nearest the origin. Among all the periods with modulus $|\omega|$ choose the one with smallest nonnegative argument and call it ω_1. (Again, such a period must exist otherwise there would be arbitrarily small nonzero periods.) If there are other periods with modulus $|\omega_1|$ besides ω_1 and $-\omega_1$, choose the one with smallest argument greater than that of ω_1 and call this ω_2. If not, find the next larger circle containing periods $\neq n\omega_1$ and choose that one of smallest nonnegative argument. Such a period exists since f has two noncollinear periods. Calling this one ω_2 we have, by construction, no periods in the triangle $0, \omega_1, \omega_2$ other than the vertices, hence the pair (ω_1, ω_2) is fundamental. □

If f and g are elliptic functions with periods ω_1 and ω_2 then their sum, difference, product and quotient are also elliptic with the same periods. So, too, is the derivative f'.

Because of periodicity, it suffices to study the behavior of an elliptic function in any period parallelogram.

Theorem 1.4. *If an elliptic function f has no poles in some period parallelogram, then f is constant.*

PROOF. If f has no poles in a period parallelogram, then f is continuous and hence bounded on the closure of the parallelogram. By periodicity, f is bounded in the whole plane. Hence, by Liouville's theorem, f is constant. □

Theorem 1.5. *If an elliptic function f has no zeros in some period parallelogram, then f is constant.*

PROOF. Apply Theorem 1.4 to the reciprocal $1/f$. □

Note. Sometimes it is inconvenient to have zeros or poles on the boundary of a period parallelogram. Since a meromorphic function has only a finite number of zeros or poles in any bounded portion of the plane, a period parallelogram can always be translated to a congruent parallelogram with no zeros or poles on its boundary. Such a translated parallelogram, with no zeros or poles on its boundary, will be called a *cell*. Its vertices need not be periods.

Theorem 1.6. *The contour integral of an elliptic function taken along the boundary of any cell is zero.*

PROOF. The integrals along parallel edges cancel because of periodicity. □

Theorem 1.7. *The sum of the residues of an elliptic function at its poles in any period parallelogram is zero.*

PROOF. Apply Cauchy's residue theorem to a cell and use Theorem 1.6. □

Note. Theorem 1.7 shows that an elliptic function which is not constant has at least two simple poles or at least one double pole in each period parallelogram.

Theorem 1.8. *The number of zeros of an elliptic function in any period parallelogram is equal to the number of poles, each counted with multiplicity.*

PROOF. The integral

$$\frac{1}{2\pi i} \int_C \frac{f'(z)}{f(z)}\, dz,$$

taken around the boundary C of a cell, counts the difference between the number of zeros and the number of poles inside the cell. But f'/f is elliptic with the same periods as f, and Theorem 1.6 tells us that this integral is zero. □

Note. The number of zeros (or poles) of an elliptic function in any period parallelogram is called the *order* of the function. Every nonconstant elliptic function has order ≥ 2.

1.5 Construction of elliptic functions

We turn now to the problem of constructing a nonconstant elliptic function. We prescribe the periods and try to find the simplest elliptic function having these periods. Since the order of such a function is at least 2 we need a second order pole or two simple poles in each period parallelogram. The two possibilities lead to two theories of elliptic functions, one developed by Weierstrass, the other by Jacobi. We shall follow Weierstrass, whose point of departure is the construction of an elliptic function with a pole of order 2 at $z = 0$ and hence at every period. Near each period ω the principal part of the Laurent expansion must have the form

$$\frac{A}{(z - \omega)^2} + \frac{B}{z - \omega}.$$

For simplicity we take $A = 1, B = 0$. Since we want such an expansion near each period ω it is natural to consider a sum of terms of this type,

$$\sum_{\omega} \frac{1}{(z - \omega)^2}$$

summed over all the periods $\omega = m\omega_1 + n\omega_2$. For fixed $z \neq \omega$ this is a double series, summed over m and n. The next two lemmas deal with convergence properties of double series of this type. In these lemmas we denote by Ω the set of all linear combinations $m\omega_1 + n\omega_2$, where m and n are arbitrary integers.

Lemma 1. *If α is real the infinite series*

$$\sum_{\substack{\omega \in \Omega \\ \omega \neq 0}} \frac{1}{\omega^{\alpha}}$$

converges absolutely if, and only if, $\alpha > 2$.

PROOF. Refer to Figure 1.3 and let r and R denote, respectively, the minimum and maximum distances from 0 to the parallelogram shown. If ω is any of the 8 nonzero periods shown in this diagram we have

$$r \leq |\omega| \leq R \qquad \text{(for 8 periods } \omega\text{)}.$$

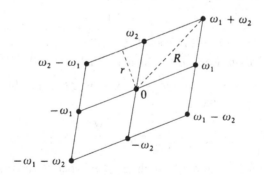

Figure 1.3

In the next concentric layer of periods surrounding these 8 we have $2 \cdot 8 = 16$ new periods satisfying the inequalities

$$2r \leq |\omega| \leq 2R \qquad \text{(for 16 new periods } \omega\text{)}.$$

In the next layer we have $3 \cdot 8 = 24$ new periods satisfying

$$3r \leq |\omega| \leq 3R \qquad \text{(for 24 new periods } \omega\text{)},$$

and so on. Therefore, we have the inequalities

$$\frac{1}{R^\alpha} \le \frac{1}{|\omega|^\alpha} \le \frac{1}{r^\alpha} \text{ for the first 8 periods } \omega,$$

$$\frac{1}{(2R)^\alpha} \le \frac{1}{|\omega|^\alpha} \le \frac{1}{(2r)^\alpha} \text{ for the next 16 periods } \omega,$$

and so on. Thus the sum $S(n) = \sum |\omega|^{-\alpha}$, taken over the $8(1 + 2 + \cdots + n)$ nonzero periods nearest the origin, satisfies the inequalities

$$\frac{8}{R^\alpha} + \frac{2 \cdot 8}{(2R)^\alpha} + \cdots + \frac{n \cdot 8}{(nR)^\alpha} \le S(n) \le \frac{8}{r^\alpha} + \frac{2 \cdot 8}{(2r)^\alpha} + \cdots + \frac{n \cdot 8}{(nr)^\alpha},$$

or

$$\frac{8}{R^\alpha} \sum_{k=1}^{n} \frac{1}{k^{\alpha-1}} \le S(n) \le \frac{8}{r^\alpha} \sum_{k=1}^{n} \frac{1}{k^{\alpha-1}}.$$

This shows that the partial sums $S(n)$ are bounded above by $8\zeta(\alpha - 1)/r^\alpha$ if $\alpha > 2$. But any partial sum lies between two such partial sums, so all of the partial sums of the series $\sum |\omega|^{-\alpha}$ are bounded above and hence the series converges if $\alpha > 2$. The lower bound for $S(n)$ also shows that the series diverges if $\alpha \le 2$. $\qquad \square$

Lemma 2. *If $\alpha > 2$ and $R > 0$ the series*

$$\sum_{|\omega| > R} \frac{1}{(z - \omega)^\alpha}$$

converges absolutely and uniformly in the disk $|z| \le R$.

PROOF. We will show that there is a constant M (depending on R and α) such that, if $\alpha \ge 1$, we have

(1)
$$\frac{1}{|z - \omega|^\alpha} \le \frac{M}{|\omega|^\alpha}$$

for all ω with $|\omega| > R$ and all z with $|z| \le R$. Then we invoke Lemma 1 to prove Lemma 2. Inequality (1) is equivalent to

(2)
$$\left| \frac{z - \omega}{\omega} \right|^\alpha \ge \frac{1}{M}.$$

To exhibit M we consider all ω in Ω with $|\omega| > R$. Choose one whose modulus is minimal, say $|\omega| = R + d$, where $d > 0$. Then if $|z| \le R$ and $|\omega| \ge R + d$ we have

$$\left| \frac{z - \omega}{\omega} \right| = \left| 1 - \frac{z}{\omega} \right| \ge 1 - \left| \frac{z}{\omega} \right| \ge 1 - \frac{R}{R + d},$$

and hence

$$\left|\frac{z - \omega}{\omega}\right|^\alpha \geq \left(1 - \frac{R}{R + d}\right)^\alpha = \frac{1}{M},$$

where

$$M = \left(1 - \frac{R}{R + d}\right)^{-\alpha}.$$

This proves (2) and also the lemma.　　　　　　　　□

As mentioned earlier, we could try to construct the simplest elliptic function by using a series of the form

$$\sum_{\omega \in \Omega} \frac{1}{(z - \omega)^2}.$$

This has the appropriate principal part near each period. However, the series does not converge absolutely so we use, instead, a series with the exponent 2 replaced by 3. This will give us an elliptic function of order 3.

Theorem 1.9. *Let f be defined by the series*

$$f(z) = \sum_{\omega \in \Omega} \frac{1}{(z - \omega)^3}.$$

Then f is an elliptic function with periods ω_1, ω_2 and with a pole of order 3 at each period ω in Ω.

PROOF. By Lemma 2 the series obtained by summing over $|\omega| > R$ converges uniformly in the disk $|z| \leq R$. Therefore it represents an analytic function in this disk. The remaining terms, which are finite in number, are also analytic in this disk except for a 3rd order pole at each period ω in the disk. This proves that f is meromorphic with a pole of order 3 at each ω in Ω.

Next we show that f has periods ω_1 and ω_2. For this we take advantage of the absolute convergence of the series. We have

$$f(z + \omega_1) = \sum_{\omega \in \Omega} \frac{1}{(z + \omega_1 - \omega)^3}.$$

But $\omega - \omega_1$ runs through all periods in Ω with ω, so the series for $f(z + \omega_1)$ is merely a rearrangement of the series for $f(z)$. By absolute convergence we have $f(z + \omega_1) = f(z)$. Similarly, $f(z + \omega_2) = f(z)$ so f is doubly periodic. This completes the proof.　　　　　　　　□

1.6 The Weierstrass $\wp$ function

Now we use the function of Theorem 1.9 to construct an elliptic function or order 2. We simply integrate the series for $f(z)$ term by term. This gives us a principal part $-(z - \omega)^{-2}/2$ near each period, so we multiply by -2 to get the principal part $(z - \omega)^{-2}$. There is also a constant of integration to reckon with. It is convenient to integrate from the origin, so we remove the term z^{-3} corresponding to $\omega = 0$, then integrate, and add the term z^{-2}. This leads us to the function

$$\frac{1}{z^2} + \int_0^z \sum_{\omega \neq 0} \frac{-2}{(t - \omega)^3} \, dt.$$

Integrating term by term we arrive at the following function, called the *Weierstrass $\wp$ function*.

Definition. The *Weierstrass $\wp$ function* is defined by the series

$$\wp(z) = \frac{1}{z^2} + \sum_{\omega \neq 0} \left\{ \frac{1}{(z - \omega)^2} - \frac{1}{\omega^2} \right\}.$$

Theorem 1.10. *The function $\wp$ so defined has periods ω_1 and ω_2. It is analytic except for a double pole at each period ω in Ω. Moreover $\wp(z)$ is an even function of z.*

PROOF. Each term in the series has modulus

$$\left| \frac{1}{(z - \omega)^2} - \frac{1}{\omega^2} \right| = \left| \frac{\omega^2 - (z - \omega)^2}{\omega^2 (z - \omega)^2} \right| = \left| \frac{z(2\omega - z)}{\omega^2 (z - \omega)^2} \right|.$$

Now consider any compact disk $|z| \leq R$. There are only a finite number of periods ω in this disk. If we exclude the terms of the series containing these periods we have, by inequality (1) obtained in the proof of Lemma 2,

$$\left| \frac{1}{(z - \omega)^2} \right| \leq \frac{M}{|\omega|^2},$$

where M is a constant depending only on R. This gives us the estimate

$$\left| \frac{z(2\omega - z)}{\omega^2 (z - \omega)^2} \right| \leq \frac{MR(2|\omega| + R)}{|\omega|^4} \leq \frac{MR(2 + R/|\omega|)}{|\omega|^3} \leq \frac{3MR}{|\omega|^3}$$

since $R < |\omega|$ for ω outside the disk $|z| \leq R$. This shows that the truncated series converges absolutely and uniformly in the disk $|z| \leq R$ and hence is analytic in this disk. The remaining terms give a second-order pole at each ω inside this disk. Therefore $\wp(z)$ is meromorphic with a pole of order 2 at each period.

Next we prove that $\wp$ is an even function. We note that

$$(-z - \omega)^2 = (z + \omega)^2 = (z - (-\omega))^2.$$

Since $-\omega$ runs through all nonzero periods with ω this shows that $\wp(-z) = \wp(z)$, so $\wp$ is even.

Finally we establish periodicity. The derivative of $\wp$ is given by

$$\wp'(z) = -2 \sum_{\omega \in \Omega} \frac{1}{(z - \omega)^3}.$$

We have already shown that this function has periods ω_1 and ω_2. Thus $\wp'(z + \omega) = \wp'(z)$ for each period ω. Therefore the function $\wp(z + \omega) - \wp(z)$ is constant. But when $z = -\omega/2$ this constant is $\wp(\omega/2) - \wp(-\omega/2) = 0$ since $\wp$ is even. Hence $\wp(z + \omega) = \wp(z)$ for each ω, so $\wp$ has the required periods. $\square$

1.7 The Laurent expansion of $\wp$ near the origin

Theorem 1.11. *Let* $r = \min \{|\omega| : \omega \neq 0\}$. *Then for* $0 < |z| < r$ *we have*

(3) $$\wp(z) = \frac{1}{z^2} + \sum_{n=1}^{\infty} (2n + 1)G_{2n+2} z^{2n},$$

where

(4) $$G_n = \sum_{\omega \neq 0} \frac{1}{\omega^n} \qquad for\ n \geq 3.$$

PROOF. If $0 < |z| < r$ then $|z/\omega| < 1$ and we have

$$\frac{1}{(z - \omega)^2} = \frac{1}{\omega^2 \left(1 - \dfrac{z}{\omega}\right)^2} = \frac{1}{\omega^2} \left(1 + \sum_{n=1}^{\infty} (n + 1)\left(\frac{z}{\omega}\right)^n\right),$$

hence

$$\frac{1}{(z - \omega)^2} - \frac{1}{\omega^2} = \sum_{n=1}^{\infty} \frac{n + 1}{\omega^{n+2}} z^n.$$

Summing over all ω we find (by absolute convergence)

$$\wp(z) = \frac{1}{z^2} + \sum_{n=1}^{\infty} (n + 1) \sum_{\omega \neq 0} \frac{1}{\omega^{n+2}} z^n = \frac{1}{z^2} + \sum_{n=1}^{\infty} (n + 1)G_{n+2} z^n,$$

where G_n is given by (4). Since $\wp(z)$ is an even function the coefficients G_{2n+1} must vanish and we obtain (3). $\square$

1.8 Differential equation satisfied by $\wp$

Theorem 1.12. *The function* $\wp$ *satisfies the nonlinear differential equation*

$$[\wp'(z)]^2 = 4\wp^3(z) - 60G_4 \wp(z) - 140G_6.$$

PROOF. We obtain this by forming a linear combination of powers of $\wp$ and $\wp'$ which eliminates the pole at $z = 0$. This gives an elliptic function which has

11

no poles and must therefore be constant. Near $z = 0$ we have

$$\wp'(z) = -\frac{2}{z^3} + 6G_4 z + 20G_6 z^3 + \cdots,$$

an elliptic function of order 3. Its square has order 6 since

$$[\wp'(z)]^2 = \frac{4}{z^6} - \frac{24G_4}{z^2} - 80G_6 + \cdots,$$

where $+ \cdots$ indicates a power series in z which vanishes at $z = 0$. Now

$$4\wp^3(z) = \frac{4}{z^6} + \frac{36G_4}{z^2} + 60G_6 + \cdots$$

hence

$$[\wp'(z)]^2 - 4\wp^3(z) = -\frac{60G_4}{z^2} - 140G_6 + \cdots$$

so

$$[\wp'(z)]^2 - 4\wp^3(z) + 60G_4 \wp(z) = -140G_6 + \cdots.$$

Since the left member has no pole at $z = 0$ it has no poles anywhere in a period parallelogram so it must be constant. Therefore this constant must be $-140G_6$ and this proves the theorem. $\qquad\square$

1.9 The Eisenstein series and the invariants g_2 and g_3

Definition. If $n \geq 3$ the series

$$G_n = \sum_{\omega \neq 0} \frac{1}{\omega^n}$$

is called the *Eisenstein series of order n*. The *invariants* g_2 and g_3 are the numbers defined by the relations

$$g_2 = 60G_4, \qquad g_3 = 140G_6.$$

The differential equation for $\wp$ now takes the form

$$[\wp'(z)]^2 = 4\wp^3(z) - g_2\wp(z) - g_3.$$

Since only g_2 and g_3 enter in the differential equation they should determine $\wp$ completely. This is actually so because all the coefficients $(2n + 1)G_{2n+2}$ in the Laurent expansion of $\wp(z)$ can be expressed in terms of g_2 and g_3.

Theorem 1.13. *Each Eisenstein series G_n is expressible as a polynomial in g_2 and g_3 with positive rational coefficients. In fact, if $b(n) = (2n + 1)G_{2n+2}$ we have the recursion relations*

$$b(1) = g_2/20, \qquad b(2) = g_3/28,$$

and

$$(2n + 3)(n - 2)b(n) = 3 \sum_{k=1}^{n-2} b(k)b(n - 1 - k) \qquad for \ n \geq 3,$$

or equivalently,

$$(2m + 1)(m - 3)(2m - 1)G_{2m} = 3 \sum_{r=2}^{m-2} (2r - 1)(2m - 2r - 1)G_{2r}G_{2m-2r}$$

for $m \geq 4$.

PROOF. Differentiation of the differential equation for $\wp$ gives another differential equation of second order satisfied by $\wp$,

(5) $$\wp''(z) = 6\wp^2(z) - \tfrac{1}{2}g_2.$$

Now we write $\wp(z) = z^{-2} + \sum_{n=1}^{\infty} b(n)z^{2n}$ and equate like powers of z in (5) to obtain the required recursion relations. $\qquad\square$

1.10 The numbers e_1, e_2, e_3

Definition. We denote by e_1, e_2, e_3 the values of $\wp$ at the half-periods,

$$e_1 = \wp\left(\frac{\omega_1}{2}\right), \qquad e_2 = \wp\left(\frac{\omega_2}{2}\right), \qquad e_3 = \wp\left(\frac{\omega_1 + \omega_2}{2}\right).$$

The next theorem shows that these numbers are the roots of the cubic polynomial $4\wp^3 - g_2\wp - g_3$.

Theorem 1.14. *We have*

$$4\wp^3(z) - g_2\wp(z) - g_3 = 4(\wp(z) - e_1)(\wp(z) - e_2)(\wp(z) - e_3).$$

Moreover, the roots e_1, e_2, e_3 are distinct, hence $g_2{}^3 - 27g_3{}^2 \neq 0$.

PROOF. Since $\wp$ is even, the derivative $\wp'$ is odd. But it is easy to show that the half-periods of an odd elliptic function are either zeros or poles. In fact, by periodicity we have $\wp'(-\tfrac{1}{2}\omega) = \wp'(\omega - \tfrac{1}{2}\omega) = \wp'(\tfrac{1}{2}\omega)$, and since $\wp'$ is odd we also have $\wp'(-\tfrac{1}{2}\omega) = -\wp'(\tfrac{1}{2}\omega)$. Hence $\wp'(\tfrac{1}{2}\omega) = 0$ if $\wp'(\tfrac{1}{2}\omega)$ is finite. Since $\wp'(z)$ has no poles at $\tfrac{1}{2}\omega_1, \tfrac{1}{2}\omega_2, \tfrac{1}{2}(\omega_1 + \omega_2)$, these points must be zeros of $\wp'$. But $\wp'$ is of order 3, so these must be simple zeros of $\wp'$. Thus $\wp'$ can have no further zeros in the period-parallelogram with vertices $0, \omega_1, \omega_2, \omega_1 + \omega_2$. The differential equation shows that each of these points is also a zero of the cubic, so we have the factorization indicated.

Next we show that the numbers e_1, e_2, e_3 are distinct. The elliptic function $\wp(z) - e_1$ vanishes at $z = \tfrac{1}{2}\omega_1$. This is a double zero since $\wp'(\tfrac{1}{2}\omega_1) = 0$. Similarly, $\wp(z) - e_2$ has a double zero at $\tfrac{1}{2}\omega_2$. If e_1 were equal to e_2, the elliptic function $\wp(z) - e_1$ would have a double zero at $\tfrac{1}{2}\omega_1$ and also a double

13

zero at $\frac{1}{2}\omega_2$, so its order would be ≥ 4. But its order is 2, so $e_1 \neq e_2$. Similarly, $e_1 \neq e_3$ and $e_2 \neq e_3$.

If a polynomial has distinct roots, its discriminant does not vanish. (See Exercise 1.7.) The discriminant of the cubic polynomial

$$4x^3 - g_2 x - g_3$$

is $g_2{}^3 - 27g_3{}^2$. When $x = \wp(z)$ the roots of this polynomial are distinct so the number $g_2{}^3 - 27g_3{}^2 \neq 0$. This completes the proof. □

1.11 The discriminant Δ

The number $\Delta = g_2{}^3 - 27g_3{}^2$ is called the *discriminant*. We regard the invariants g_2 and g_3 and the discriminant Δ as functions of the periods ω_1 and ω_2 and we write

$$g_2 = g_2(\omega_1, \omega_2), \qquad g_3 = g_3(\omega_1, \omega_2), \qquad \Delta = \Delta(\omega_1, \omega_2).$$

The Eisenstein series show that g_2 and g_3 are homogeneous functions of degrees -4 and -6, respectively. That is, we have

$$g_2(\lambda\omega_1, \lambda\omega_2) = \lambda^{-4}g_2(\omega_1, \omega_2) \quad \text{and} \quad g_3(\lambda\omega_1, \lambda\omega_2) = \lambda^{-6}g_3(\omega_1, \omega_2)$$

for any $\lambda \neq 0$. Hence Δ is homogeneous of degree -12,

$$\Delta(\lambda\omega_1, \lambda\omega_2) = \lambda^{-12}\Delta(\omega_1, \omega_2).$$

Taking $\lambda = 1/\omega_1$ and writing $\tau = \omega_2/\omega_1$ we obtain

$$g_2(1, \tau) = \omega_1{}^4 g_2(\omega_1, \omega_2), \qquad g_3(1, \tau) = \omega_1{}^6 g_3(\omega_1, \omega_2),$$
$$\Delta(1, \tau) = \omega_1{}^{12}\Delta(\omega_1, \omega_2).$$

Therefore a change of scale converts g_2, g_3 and Δ into functions of one complex variable τ. We shall label ω_1 and ω_2 in such a way that their ratio $\tau = \omega_2/\omega_1$ has positive imaginary part and study these functions in the upper half-plane $\text{Im}(\tau) > 0$. We denote the upper half-plane $\text{Im}(\tau) > 0$ by H.

If $\tau \in H$ we write $g_2(\tau)$, $g_3(\tau)$ and $\Delta(\tau)$ for $g_2(1, \tau)$ $g_3(1, \tau)$ and $\Delta(1, \tau)$, respectively. Thus, we have

$$g_2(\tau) = 60 \sum_{\substack{m,n=-\infty \\ (m,n)\neq(0,0)}}^{+\infty} \frac{1}{(m + n\tau)^4},$$

$$g_3(\tau) = 140 \sum_{\substack{m,n=-\infty \\ (m,n)\neq(0,0)}}^{+\infty} \frac{1}{(m + n\tau)^6}$$

and

$$\Delta(\tau) = g_2{}^3(\tau) - 27g_3{}^2(\tau).$$

Theorem 1.14 shows that $\Delta(\tau) \neq 0$ for all τ in H.

1.12 Klein's modular function $J(\tau)$

Klein's function is a combination of g_2 and g_3 defined in such a way that, as a function of the periods ω_1 and ω_2, it is homogeneous of degree 0.

Definition. If ω_2/ω_1 is not real we define

$$J(\omega_1, \omega_2) = \frac{g_2{}^3(\omega_1, \omega_2)}{\Delta(\omega_1, \omega_2)}.$$

Since $g_2{}^3$ and Δ are homogeneous of the same degree we have $J(\lambda\omega_1, \lambda\omega_2) = J(\omega_1, \omega_2)$. In particular, if $\tau \in H$ we have

$$J(1, \tau) = J(\omega_1, \omega_2).$$

Thus $J(\omega_1, \omega_2)$ is a function of the ratio τ alone. We write $J(\tau)$ for $J(1, \tau)$.

Theorem 1.15. *The functions $g_2(\tau)$, $g_3(\tau)$, $\Delta(\tau)$, and $J(\tau)$ are analytic in H.*

PROOF. Since $\Delta(\tau) \neq 0$ in H it suffices to prove that g_2 and g_3 are analytic in H. Both g_2 and g_3 are given by double series of the form

$$\sum_{\substack{m, n = -\infty \\ (m, n) \neq (0, 0)}}^{+\infty} \frac{1}{(m + n\tau)^\alpha}$$

·ith $\alpha > 2$. Let $\tau = x + iy$, where $y > 0$. We shall prove that if $\alpha > 2$ this series converges absolutely for any fixed τ in H and uniformly in every strip S of the form

$$S = \{x + iy : |x| \leq A, y \geq \delta > 0\}.$$

(See Figure 1.4.) To do this we prove that there is a constant $M > 0$, depending only on A and on δ, such that

(6)
$$\frac{1}{|m + n\tau|^\alpha} \leq \frac{M}{|m + ni|^\alpha}$$

for all τ in S and all $(m, n) \neq (0, 0)$. Then we invoke Lemma 1.

To prove (6) it suffices to prove that

$$|m + n\tau|^2 > K|m + ni|^2$$

for some $K > 0$ which depends only on A and δ, or that

(7)
$$(m + nx)^2 + (ny)^2 > K(m^2 + n^2).$$

If $n = 0$ this inequality holds with any K such that $0 < K < 1$. If $n \neq 0$ let $q = m/n$. Proving (7) is equivalent to showing that

(8)
$$\frac{(q + x)^2 + y^2}{1 + q^2} > K$$

15

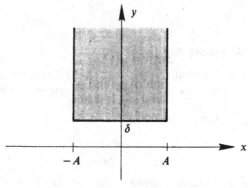

Figure 1.4

for some $K > 0$. We will prove that (8) holds for all q, with

$$K = \frac{\delta^2}{1 + (A + \delta)^2}$$

if $|x| \leq A$ and $y \geq \delta$. (This proof was suggested by Christopher Henley.)

If $|q| \leq A + \delta$ inequality (8) holds trivially since $(q + x)^2 \geq 0$ and $y^2 \geq \delta^2$. If $|q| > A + \delta$ then $|x/q| < |x|/(A + \delta) \leq A/(A + \delta) < 1$ so

$$\left| 1 + \frac{x}{q} \right| \geq 1 - \left| \frac{x}{q} \right| > 1 - \frac{A}{A + \delta} = \frac{\delta}{A + \delta}$$

hence

$$|q + x| \geq \frac{q\delta}{A + \delta}$$

and

(9) $$\frac{(q + x)^2 + y^2}{1 + q^2} > \frac{\delta^2}{(A + \delta)^2} \frac{q^2}{1 + q^2}.$$

Now $q^2/(1 + q^2)$ is an increasing function of q^2 so

$$\frac{q^2}{1 + q^2} \geq \frac{(A + \delta)^2}{1 + (A + \delta)^2}$$

when $q^2 > (A + \delta)^2$. Using this in (9) we obtain (8) with the specified K. $\square$

1.13 Invariance of J under unimodular transformations

If ω_1, ω_2 are given periods with nonreal ratio, introduce new periods $\omega_1{}'$, $\omega_2{}'$ by the relations

$$\omega_2{}' = a\omega_2 + b\omega_1, \qquad \omega_1{}' = c\omega_2 + d\omega_1,$$

where a, b, c, d are integers such that $ad - bc = 1$. Then the pair (ω_1', ω_2') is equivalent to (ω_1, ω_2); that is, it generates the same set of periods Ω. Therefore $g_2(\omega_1', \omega_2') = g_2(\omega_1, \omega_2)$ and $g_3(\omega_1', \omega_2') = g_3(\omega_1, \omega_2)$ since g_2 and g_3 depend only on the set of periods Ω. Consequently, $\Delta(\omega_1', \omega_2') = \Delta(\omega_1, \omega_2)$ and $J(\omega_1', \omega_2') = J(\omega_1, \omega_2)$.

The ratio of the new periods is

$$\tau' = \frac{\omega_2'}{\omega_1'} = \frac{a\omega_2 + b\omega_1}{c\omega_2 + d\omega_1} = \frac{a\tau + b}{c\tau + d},$$

where $\tau = \omega_2/\omega_1$. An easy calculation shows that

$$\text{Im}(\tau') = \text{Im}\left(\frac{a\tau + b}{c\tau + d}\right) = \frac{ad - bc}{|c\tau + d|^2} \text{Im}(\tau) = \frac{\text{Im}(\tau)}{|c\tau + d|^2}.$$

Hence $\tau' \in H$ if and only if $\tau \in H$. The equation

$$\tau' = \frac{a\tau + b}{c\tau + d}$$

is called a *unimodular transformation* if a, b, c, d are integers with $ad - bc = 1$. The set of all unimodular transformations forms a group (under composition) called the *modular group*. This group will be discussed further in the next chapter. The foregoing remarks show that the function $J(\tau)$ is invariant under the transformations of the modular group. That is, we have:

Theorem 1.16. *If $\tau \in H$ and a, b, c, d are integers with $ad - bc = 1$, then $(a\tau + b)/(c\tau + d) \in H$ and*

(10)
$$J\left(\frac{a\tau + b}{c\tau + d}\right) = J(\tau).$$

Note. A particular unimodular transformation is $\tau' = \tau + 1$, hence (10) shows that $J(\tau + 1) = J(\tau)$. In other words, $J(\tau)$ is a periodic function of τ with period 1. The next theorem shows that $J(\tau)$ has a Fourier expansion.

Theorem 1.17. *If $\tau \in H$, $J(\tau)$ can be represented by an absolutely convergent Fourier series*

(11)
$$J(\tau) = \sum_{n=-\infty}^{\infty} a(n)e^{2\pi in\tau}.$$

PROOF. Introduce the change of variable

$$x = e^{2\pi i\tau}.$$

Then the upper half-plane H maps into the punctured unit disk

$$D = \{x : 0 < |x| < 1\}.$$

(See Figure 1.5.) Each τ in H maps onto a unique point x in D, but each x in D is the image of infinitely many points in H. If τ and τ' map onto x then $e^{2\pi i\tau} = e^{2\pi i\tau'}$ so τ and τ' differ by an integer.

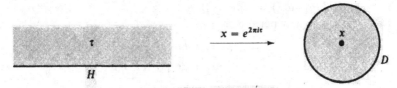

Figure 1.5

If $x \in D$, let

$$f(x) = J(\tau)$$

where τ is any of the points in H which map onto x. Since J is periodic with period 1, J has the same value at all these points so $f(x)$ is well-defined. Now f is analytic in D because

$$f'(x) = \frac{d}{dx} J(\tau) = \frac{d}{d\tau} J(\tau) \frac{d\tau}{dx} = J'(\tau) \left/ \frac{dx}{d\tau} \right. = \frac{J'(\tau)}{2\pi i e^{2\pi i\tau}},$$

so $f'(x)$ exists at each point in D. Since f is analytic in D it has a Laurent expansion about 0,

$$f(x) = \sum_{n=-\infty}^{\infty} a(n)x^n,$$

absolutely convergent for each x in D. Replacing x by $e^{2\pi i\tau}$ we see that $J(\tau)$ has the absolutely convergent Fourier expansion in (11). $\qquad \square$

Later we will show that $a_{-n} = 0$ for $n \geq 2$, that $a_{-1} = 12^{-3}$, and that the Fourier expansion of $12^3 J(\tau)$ has integer coefficients. To do this we first determine the Fourier expansions of $g_2(\tau)$, $g_3(\tau)$ and $\Delta(\tau)$.

1.14 The Fourier expansions of $g_2(\tau)$ and $g_3(\tau)$

Each Eisenstein series $\sum_{(m, n) \neq (0, 0)} (m + n\tau)^{-k}$ is a periodic function of τ of period 1. In particular, $g_2(\tau)$ and $g_3(\tau)$ are periodic with period 1. In this section we determine their Fourier coefficients explicitly.

We recall that

$$g_2(\tau) = 60 \sum_{(m, n) \neq (0, 0)} \frac{1}{(m + n\tau)^4}, \qquad g_3(\tau) = 140 \sum_{(m, n) \neq (0, 0)} \frac{1}{(m + n\tau)^6}.$$

These are double series in m and n. First we obtain Fourier expansions for the simpler series

$$\sum_{m=-\infty}^{+\infty} \frac{1}{(m+n\tau)^4} \quad \text{and} \quad \sum_{m=-\infty}^{+\infty} \frac{1}{(m+n\tau)^6}.$$

Lemma 3. *If $\tau \in H$ and $n > 0$ we have the Fourier expansions*

$$\sum_{m=-\infty}^{+\infty} \frac{1}{(m+n\tau)^4} = \frac{8\pi^4}{3} \sum_{r=1}^{\infty} r^3 e^{2\pi i r n\tau}$$

and

$$\sum_{m=-\infty}^{+\infty} \frac{1}{(m+n\tau)^6} = -\frac{8\pi^6}{15} \sum_{r=1}^{\infty} r^5 e^{2\pi i r n\tau}.$$

PROOF. Start with the partial fraction decomposition of the cotangent:

$$\pi \cot \pi\tau = \frac{1}{\tau} + \sum_{\substack{m=-\infty \\ m\neq 0}}^{+\infty} \left(\frac{1}{\tau+m} - \frac{1}{m}\right).$$

Let $x = e^{2\pi i \tau}$. If $\tau \in H$ then $|x| < 1$ and we find

$$\pi \cot \pi\tau = \pi \frac{\cos \pi\tau}{\sin \pi\tau} = \pi i \frac{e^{2\pi i\tau}+1}{e^{2\pi i\tau}-1} = \pi i \frac{x+1}{x-1} = -\pi i\left(\frac{x}{1-x} + \frac{1}{1-x}\right)$$

$$= -\pi i\left(\sum_{r=1}^{\infty} x^r + \sum_{r=0}^{\infty} x^r\right) = -\pi i\left(1 + 2\sum_{r=1}^{\infty} x^r\right).$$

In other words, if $\tau \in H$ we have

$$\frac{1}{\tau} + \sum_{\substack{m=-\infty \\ m\neq 0}}^{+\infty} \left(\frac{1}{\tau+m} - \frac{1}{m}\right) = -\pi i\left(1 + 2\sum_{r=1}^{\infty} e^{2\pi i r\tau}\right).$$

Differentiating repeatedly we find

(12)
$$-\frac{1}{\tau^2} - \sum_{\substack{m=-\infty \\ m\neq 0}}^{\infty} \frac{1}{(\tau+m)^2} = -(2\pi i)^2 \sum_{r=1}^{\infty} r e^{2\pi i r\tau}$$

$$-3! \sum_{m=-\infty}^{+\infty} \frac{1}{(\tau+m)^4} = -(2\pi i)^4 \sum_{r=1}^{\infty} r^3 e^{2\pi i r\tau}$$

and

$$-5! \sum_{m=-\infty}^{+\infty} \frac{1}{(\tau+m)^6} = -(2\pi i)^6 \sum_{r=1}^{\infty} r^5 e^{2\pi i r\tau}.$$

Replacing τ by $n\tau$ we obtain Lemma 3. $\qquad \square$

Theorem 1.18. *If $\tau \in H$ we have the Fourier expansions*

$$g_2(\tau) = \frac{4\pi^4}{3}\left\{1 + 240 \sum_{k=1}^{\infty} \sigma_3(k)e^{2\pi ik\tau}\right\}$$

and

$$g_3(\tau) = \frac{8\pi^6}{27}\left\{1 - 504 \sum_{k=1}^{\infty} \sigma_5(k)e^{2\pi ik\tau}\right\},$$

where $\sigma_\alpha(k) = \sum_{d|k} d^\alpha$.

PROOF. We write

$$g_2(\tau) = 60 \sum_{\substack{m,n=-\infty \\ (m,n)\neq(0,0)}}^{+\infty} \frac{1}{(m+n\tau)^4}$$

$$= 60\left\{\sum_{\substack{m=-\infty \\ m\neq 0(n=0)}}^{+\infty} \frac{1}{m^4} + \sum_{n=1}^{\infty}\sum_{m=-\infty}^{+\infty}\left(\frac{1}{(m+n\tau)^4} + \frac{1}{(m-n\tau)^4}\right)\right\}$$

$$= 60\left\{2\zeta(4) + 2\sum_{n=1}^{\infty}\sum_{m=-\infty}^{+\infty} \frac{1}{(m+n\tau)^4}\right\}$$

$$= 60\left\{\frac{2\pi^4}{90} + \frac{16\pi^4}{3}\sum_{n=1}^{\infty}\sum_{r=1}^{\infty} r^3 x^{nr}\right\}$$

where $x = e^{2\pi i\tau}$. In the last double sum we collect together those terms for which nr is constant and we obtain the expansion for $g_2(\tau)$. The formula for $g_3(\tau)$ is similarly proved. $\square$

1.15 The Fourier expansions of $\Delta(\tau)$ and $J(\tau)$

Theorem 1.19. *If $\tau \in H$ we have the Fourier expansion*

$$\Delta(\tau) = (2\pi)^{12} \sum_{n=1}^{\infty} \tau(n)e^{2\pi in\tau}$$

where the coefficients $\tau(n)$ are integers, with $\tau(1) = 1$ and $\tau(2) = -24$.

Note. The arithmetical function $\tau(n)$ is called *Ramanujan's tau function.* Some of its arithmetical properties are described in Chapter 4.

PROOF. Let

$$x = e^{2\pi i\tau}, \qquad A = \sum_{n=1}^{\infty} \sigma_3(n)x^n, \qquad B = \sum_{n=1}^{\infty} \sigma_5(n)x^n.$$

Then

$$\Delta(\tau) = g_2{}^3(\tau) - 27g_3{}^2(\tau) = \frac{64\pi^{12}}{27}\{(1 + 240A)^3 - (1 - 504B)^2\}.$$

Now A and B have integer coefficients, and

$$
\begin{aligned}
(1 + 240A)^3 - (1 - 504B)^2 &= 1 + 720A + 3(240)^2 A^2 + (240)^3 A^3 - 1 \\
&\quad + 1008B - (504)^2 B^2 \\
&= 12^2(5A + 7B) \\
&\quad + 12^3(100A^2 - 147B^2 + 8000A^3).
\end{aligned}
$$

But

$$
5A + 7B = \sum_{n=1}^{\infty} \{5\sigma_3(n) + 7\sigma_5(n)\}x^n
$$

and

$$
5d^3 + 7d^5 = d^3(5 + 7d^2) \equiv \begin{cases} d^3(d^2 - 1) \equiv 0 \pmod 3 \\ d^3(1 - d^2) \equiv 0 \pmod 4 \end{cases}
$$

so

$$
5d^3 + 7d^5 \equiv 0 \pmod{12}.
$$

Hence 12^3 is a factor of each coefficient in the power series expansion of $(1 + 240A)^3 - (1 - 504B)^2$ so

$$
\Delta(\tau) = \frac{64\pi^{12}}{27}\left\{12^3 \sum_{n=1}^{\infty} \tau(n)e^{2\pi i n \tau}\right\} = (2\pi)^{12} \sum_{n=1}^{\infty} \tau(n)e^{2\pi i n \tau}
$$

where the $\tau(n)$ are integers. The coefficient of x is $12^2(5 + 7)$, so $\tau(1) = 1$. Similarly, we find $\tau(2) = -24$. $\qquad\square$

Theorem 1.20. *If $\tau \in H$ we have the Fourier expansion*

$$
12^3 J(\tau) = e^{-2\pi i \tau} + 744 + \sum_{n=1}^{\infty} c(n)e^{2\pi i n \tau},
$$

where the $c(n)$ are integers.

PROOF. We agree to write I for any power series in x with integer coefficients. Then if $x = e^{2\pi i \tau}$ we have

$$
g_2^{\ 3}(\tau) = \tfrac{64}{27}\pi^{12}(1 + 240x + I)^3 = \tfrac{64}{27}\pi^{12}(1 + 720x + I),
$$
$$
\Delta(\tau) = \tfrac{64}{27}\pi^{12}\{12^3 x(1 - 24x + I)\}
$$

and hence

$$
J(\tau) = \frac{g_2^{\ 3}(\tau)}{\Delta(\tau)} = \frac{1 + 720x + I}{12^3 x(1 - 24x + I)} = \frac{1}{12^3 x}(1 + 720x + I)(1 + 24x + I)
$$

so

$$
12^3 J(\tau) = \frac{1}{x} + 744 + \sum_{n=1}^{\infty} c(n)x^n,
$$

where the $c(n)$ are integers. $\qquad\square$

Note. The coefficients $c(n)$ have been calculated for $n \leq 100$. Berwick calculated the first 7 in 1916, Zuckerman the first 24 in 1939, and Van Wijngaarden the first 100 in 1953. The first few are repeated here.

$c(0) = 744$
$c(1) = 196, 884$
$c(2) = 21, 493, 760$
$c(3) = 864, 299, 970$
$c(4) = 20, 245, 856, 256$
$c(5) = 333, 202, 640, 600$
$c(6) = 4, 252, 023, 300, 096$
$c(7) = 44, 656, 994, 071, 935$
$c(8) = 401, 490, 886, 656, 000$

The integers $c(n)$ have a number of interesting arithmetical properties. In 1942 D. H. Lehmer [20] proved that

$$(n + 1)c(n) \equiv 0 \pmod{24} \quad \text{for all } n \geq 1.$$

In 1949 Joseph Lehner [23] discovered divisibility properties of a different kind. For example, he proved that

$$c(5n) \equiv 0 \pmod{25},$$
$$c(7n) \equiv 0 \pmod{7},$$
$$c(11n) \equiv 0 \pmod{11}.$$

He also discovered congruences for higher powers of 5, 7, 11 and, in a later paper [24] found similar results for the primes 2 and 3. In Chapter 4 we will describe how some of Lehner's congruences are obtained.

An asymptotic formula for $c(n)$ was discovered by Petersson [31] in 1932. It states that

$$c(n) \sim \frac{e^{4\pi\sqrt{n}}}{\sqrt{2}\, n^{3/4}} \text{ as } n \to \infty.$$

This formula was rediscovered independently by Rademacher [37] in 1938.

The coefficients $\tau(n)$ in the Fourier expansion of $\Delta(\tau)$ have also been extensively tabulated by D. H. Lehmer [19] and others. The first ten entries in Lehmer's table are repeated here:

$\tau(1) = 1$	$\tau(6) = -6048$
$\tau(2) = -24$	$\tau(7) = -16744$
$\tau(3) = 252$	$\tau(8) = 84480$
$\tau(4) = -1472$	$\tau(9) = -113643$
$\tau(5) = 4830$	$\tau(10) = -115920.$

Lehmer has conjectured that $\tau(n) \neq 0$ for all n and has verified this for all $n < 214928639999$ by studying various congruences satisfied by $\tau(n)$. For papers on $\tau(n)$ see Section F35 of [27].

Exercises for Chapter 1

1. Given two pairs of complex numbers (ω_1, ω_2) and (ω_1', ω_2') with nonreal ratios ω_2/ω_1 and ω_2'/ω_1'. Prove that they generate the same set of periods if, and only if, there is a 2×2 matrix $\begin{pmatrix} a & b \\ c & d \end{pmatrix}$ with integer entries and determinant ± 1 such that

$$\begin{pmatrix} \omega_2' \\ \omega_1' \end{pmatrix} = \begin{pmatrix} a & b \\ c & d \end{pmatrix} \begin{pmatrix} \omega_2 \\ \omega_1 \end{pmatrix}.$$

2. Let $S(0)$ denote the sum of the zeros of an elliptic function f in a period parallelogram, and let $S(\infty)$ denote the sum of the poles in the same parallelogram. Prove that $S(0) - S(\infty)$ is a period of f. [*Hint*: Integrate $zf''(z)/f(z)$.]

3. (a) Prove that $\wp(u) = \wp(v)$ if, and only if, $u - v$ or $u + v$ is a period of $\wp$.
(b) Let $a_1, \ldots, a_n$ and $b_1, \ldots, b_m$ be complex numbers such that none of the numbers $\wp(a_i) - \wp(b_j)$ is zero. Let

$$f(z) = \prod_{k=1}^{n} [\wp(z) - \wp(a_k)] \bigg/ \prod_{r=1}^{m} [\wp(z) - \wp(b_r)].$$

Prove that f is an even elliptic function with zeros at $a_1, \ldots, a_n$ and poles at $b_1, \ldots, b_m$.

4. Prove that every even elliptic function f is a rational function of $\wp$, where the periods of $\wp$ are a subset of the periods of f.

5. Prove that every elliptic function f can be expressed in the form

$$f(z) = R_1[\wp(z)] + \wp'(z)R_2[\wp(z)]$$

where R_1 and R_2 are rational functions and $\wp$ has the same set of periods as f.

6. Let f and g be two elliptic functions with the same set of periods. Prove that there exists a polynomial $P(x, y)$, not identically zero, such that

$$P[f(z), g(z)] = C$$

where C is a constant (depending on f and g but not on z).

7. The discriminant of the polynomial $f(x) = 4(x - x_1)(x - x_2)(x - x_3)$ is the product $16\{(x_2 - x_1)(x_3 - x_2)(x_3 - x_1)\}^2$. Prove that the discriminant of $f(x) = 4x^3 - ax - b$ is $a^3 - 27b^2$.

8. The differential equation for $\wp$ shows that $\wp'(z) = 0$ if $z = \omega_1/2$, $\omega_2/2$ or $(\omega_1 + \omega_2)/2$. Show that

$$\wp''\left(\frac{\omega_1}{2}\right) = 2(e_1 - e_2)(e_1 - e_3)$$

and obtain corresponding formulas for $\wp''(\omega_2/2)$ and $\wp''((\omega_1 + \omega_2)/2)$.

9. According to Exercise 4, the function $\wp(2z)$ is a rational function of $\wp(z)$. Prove that, in fact,

$$\wp(2z) = \frac{\{\wp^2(z) + \frac{1}{4}g_2\}^2 + 2g_3\,\wp(z)}{4\wp^3(z) - g_2\wp(z) - g_3} = -2\wp(z) + \frac{1}{4}\left(\frac{\wp''(z)}{\wp'(z)}\right)^2.$$

10. Let ω_1 and ω_2 be complex numbers with nonreal ratio. Let $f(z)$ be an entire function and assume there are constants a and b such that

$$f(z + \omega_1) = af(z), \qquad f(z + \omega_2) = bf(z),$$

for all z. Prove that $f(z) = Ae^{Bz}$, where A and B are constants.

11. If $k \geq 2$ and $\tau \in H$ prove that the Eisenstein series

$$G_{2k}(\tau) = \sum_{(m,\,n) \neq (0,\,0)} (m + n\tau)^{-2k}$$

has the Fourier expansion

$$G_{2k}(\tau) = 2\zeta(2k) + \frac{2(2\pi i)^{2k}}{(2k-1)!} \sum_{n=1}^{\infty} \sigma_{2k-1}(n)e^{2\pi i n\tau}.$$

12. Refer to Exercise 11. If $\tau \in H$ prove that

$$G_{2k}(-1/\tau) = \tau^{2k}G_{2k}(\tau)$$

and deduce that

$$\begin{aligned} G_{2k}(i/2) &= (-4)^k G_{2k}(2i) \quad \text{for all } k \geq 2, \\ G_{2k}(i) &= 0 \qquad\qquad\quad \text{if } k \text{ is odd,} \\ G_{2k}(e^{2\pi i/3}) &= 0 \qquad\qquad\quad \text{if } k \not\equiv 0 \ (\text{mod } 3). \end{aligned}$$

13. Ramanujan's tau function $\tau(n)$ is defined by the Fourier expansion

$$\Delta(\tau) = (2\pi)^{12} \sum_{n=1}^{\infty} \tau(n)e^{2\pi i n\tau},$$

derived in Theorem 1.19. Prove that

$$\tau(n) = 8000\{(\sigma_3 \circ \sigma_3) \circ \sigma_3\}(n) - 147(\sigma_5 \circ \sigma_5)(n),$$

where $f \circ g$ denotes the Cauchy product of two sequences,

$$(f \circ g)(n) = \sum_{k=0}^{n} f(k)g(n-k),$$

and $\sigma_\alpha(n) = \sum_{d|n} d^\alpha$ for $n \geq 1$, with $\sigma_3(0) = \frac{1}{240}$, $\sigma_5(0) = -\frac{1}{504}$.
[*Hint*: Theorem 1.18.]

14. A series of the form $\sum_{n=1}^{\infty} f(n)x^n/(1-x^n)$ is called a *Lambert series*. Assuming absolute convergence, prove that

$$\sum_{n=1}^{\infty} f(n)\frac{x^n}{1-x^n} = \sum_{n=1}^{\infty} F(n)x^n,$$

where

$$F(n) = \sum_{d|n} f(d).$$

Apply this result to obtain the following formulas, valid for $|x| < 1$.

(a) $\displaystyle\sum_{n=1}^{\infty} \frac{\mu(n)x^n}{1 - x^n} = x.$

(b) $\displaystyle\sum_{n=1}^{\infty} \frac{\varphi(n)x^n}{1 - x^n} = \frac{x}{(1 - x)^2}.$

(c) $\displaystyle\sum_{n=1}^{\infty} \frac{n^\alpha x^n}{1 - x^n} = \sum_{n=1}^{\infty} \sigma_\alpha(n)x^n.$

(d) $\displaystyle\sum_{n=1}^{\infty} \frac{\lambda(n)x^n}{1 - x^n} = \sum_{n=1}^{\infty} x^{n^2}.$

(e) Use the result in (c) to express $g_2(\tau)$ and $g_3(\tau)$ in terms of Lambert series in $x = e^{2\pi i \tau}$.

Note. In (a), $\mu(n)$ is the Möbius function; in (b), $\varphi(n)$ is Euler's totient; and in (d), $\lambda(n)$ is Liouville's function.

15. Let

$$G(x) = \sum_{n=1}^{\infty} \frac{n^5 x^n}{1 - x^n},$$

and let

$$F(x) = \sum_{\substack{n=1 \\ (n\ \mathrm{odd})}}^{\infty} \frac{n^5 x^n}{1 + x^n}.$$

(a) Prove that $F(x) = G(x) - 34G(x^2) + 64G(x^4)$.

(b) Prove that

$$\sum_{\substack{n=1 \\ (n\ \mathrm{odd})}}^{\infty} \frac{n^5}{1 + e^{n\pi}} = \frac{31}{504}.$$

(c) Use Theorem 12.17 in [4] to prove the more general result

$$\sum_{\substack{n=1 \\ (n\ \mathrm{odd})}}^{\infty} \frac{n^{4k+1}}{1 + e^{n\pi}} = \frac{2^{4k+1} - 1}{8k + 4} B_{4k+2}.$$

2

The modular group and modular functions

2.1 Möbius transformations

In the foregoing chapter we encountered unimodular transformations

$$\tau' = \frac{a\tau + b}{c\tau + d}$$

where a, b, c, d are integers with $ad - bc = 1$. This chapter studies such transformations in greater detail and also studies functions which, like $J(\tau)$, are invariant under unimodular transformations. We begin with some remarks concerning the more general transformations

$$(1) \qquad f(z) = \frac{az + b}{cz + d}$$

where a, b, c, d are arbitrary complex numbers.

Equation (1) defines $f(z)$ for all z in the extended complex number system $C^* = C \cup \{\infty\}$ except for $z = -d/c$ and $z = \infty$. We extend the definition of f to all of C^* by defining

$$f\left(\frac{-d}{c}\right) = \infty \qquad \text{and} \qquad f(\infty) = \frac{a}{c},$$

with the usual convention that $z/0 = \infty$ if $z \neq 0$.

First we note that

$$(2) \qquad f(w) - f(z) = \frac{(ad - bc)(w - z)}{(cw + d)(cz + d)},$$

which shows that f is constant if $ad - bc = 0$. To avoid this degenerate case we assume that $ad - bc \neq 0$. The resulting rational function is called a

Möbius transformation. It is analytic everywhere on C^* except for a simple pole at $z = -d/c$.

Equation (2) shows that every Möbius transformation is one-to-one on C^*. Solving (1) for z in terms of $f(z)$ we find

$$z = \frac{df(z) - b}{-cf(z) + a},$$

so f maps C^* onto C^*. This also shows that the inverse function f^{-1} is a Möbius transformation.

Dividing by $w - z$ in (2) and letting $w \to z$ we obtain

$$f'(z) = \frac{ad - bc}{(cz + d)^2},$$

hence $f'(z) \neq 0$ at each point of analyticity. Therefore f is conformal everywhere except possibly at the pole $z = -d/c$.

Möbius transformations map circles onto circles (with straight lines being considered as special cases of circles). To prove this we consider the equation

$$(3) \qquad Az\bar{z} + Bz + \bar{B}\bar{z} + C = 0,$$

where A and C are real. The points on any circle satisfy such an equation with $A \neq 0$, and the points on any line satisfy such an equation with $A = 0$. Replacing z in (3) by $(aw + b)/(cw + d)$ we find that w satisfies an equation of the same type,

$$A'w\bar{w} + B'w + \bar{B}'\bar{w} + C' = 0$$

where A' and C' are also real. Hence every Möbius transformation maps a circle or straight line onto a circle or straight line.

A Möbius transformation remains unchanged if we multiply all the coefficients a, b, c, d by the same nonzero constant. Therefore there is no loss in generality in assuming that $ad - bc = 1$.

For each Möbius transformation (1) with $ad - bc = 1$ we associate the 2×2 matrix

$$A = \begin{pmatrix} a & b \\ c & d \end{pmatrix}.$$

Then $\det A = ad - bc = 1$. If A and B are the matrices associated with Möbius transformations f and g, respectively, then it is easy to verify that the matrix product AB is associated with the composition $f \circ g$, where $(f \circ g)(z) = f(g(z))$. The identity matrix $I = \begin{pmatrix} 1 & 0 \\ 0 & 1 \end{pmatrix}$ is associated with the identity transformation

$$f(z) = z = \frac{1z + 0}{0z + 1},$$

and the matrix inverse

$$A^{-1} = \begin{pmatrix} d & -b \\ -c & a \end{pmatrix}$$

is associated with the inverse of f,

$$f^{-1}(z) = \frac{dz - b}{-cz + a}.$$

Thus we see that the set of all Möbius transformations with $ad - bc = 1$ forms a group under composition. This chapter is concerned with an important subgroup in which the coefficients a, b, c, d are integers.

2.2 The modular group Γ

The set of all Möbius transformations of the form

$$\tau' = \frac{a\tau + b}{c\tau + d},$$

where a, b, c, d are integers with $ad - bc = 1$, is called the *modular group* and is denoted by Γ. The group can be represented by 2×2 integer matrices

$$A = \begin{pmatrix} a & b \\ c & d \end{pmatrix} \quad \text{with det } A = 1,$$

provided we identify each matrix with its negative, since A and $-A$ represent the same transformation. Ordinarily we will make no distinction between the matrix and the transformation. If $A = \begin{pmatrix} a & b \\ c & d \end{pmatrix}$ we write

$$A\tau = \frac{a\tau + b}{c\tau + d}.$$

The first theorem shows that Γ is generated by two transformations,

$$T\tau = \tau + 1 \quad \text{and} \quad S\tau = -\frac{1}{\tau}.$$

Theorem 2.1. *The modular group Γ is generated by the two matrices*

$$T = \begin{pmatrix} 1 & 1 \\ 0 & 1 \end{pmatrix} \quad \text{and} \quad S = \begin{pmatrix} 0 & -1 \\ 1 & 0 \end{pmatrix}.$$

That is, every A in Γ can be expressed in the form

$$A = T^{n_1}ST^{n_2}S \cdots ST^{n_k}$$

where the n_i are integers. This representation is not unique.

PROOF. Consider first a particular example, say

$$A = \begin{pmatrix} 4 & 9 \\ 11 & 25 \end{pmatrix}.$$

We will express A as a product of powers of S and T. Since $S^2 = I$, only the first power of S will occur.

Consider the matrix product

$$AT^n = \begin{pmatrix} 4 & 9 \\ 11 & 25 \end{pmatrix}\begin{pmatrix} 1 & n \\ 0 & 1 \end{pmatrix} = \begin{pmatrix} 4 & 4n + 9 \\ 11 & 11n + 25 \end{pmatrix}.$$

Note that the first column remains unchanged. By a suitable choice of n we can make $|11n + 25| < 11$. For example, taking $n = -2$ we find $11n + 25 = 3$ and

$$AT^{-2} = \begin{pmatrix} 4 & 1 \\ 11 & 3 \end{pmatrix}.$$

Thus by multiplying A by a suitable power of T we get a matrix $\begin{pmatrix} a & b \\ c & d \end{pmatrix}$ with $|d| < |c|$. Next, multiply by S on the right:

$$AT^{-2}S = \begin{pmatrix} 4 & 1 \\ 11 & 3 \end{pmatrix}\begin{pmatrix} 0 & -1 \\ 1 & 0 \end{pmatrix} = \begin{pmatrix} 1 & -4 \\ 3 & -11 \end{pmatrix}.$$

This interchanges the two columns and changes the sign of the second column. Again, multiplication by a suitable power of T gives us a matrix with $|d| < |c|$. In this case we can use either T^4 or T^3. Choosing T^4 we find

$$AT^{-2}ST^4 = \begin{pmatrix} 1 & -4 \\ 3 & -11 \end{pmatrix}\begin{pmatrix} 1 & 4 \\ 0 & 1 \end{pmatrix} = \begin{pmatrix} 1 & 0 \\ 3 & 1 \end{pmatrix}.$$

Multiplication by S gives

$$AT^{-2}ST^4S = \begin{pmatrix} 0 & -1 \\ 1 & -3 \end{pmatrix}.$$

Now we multiply by T^3 to get

$$AT^{-2}ST^4ST^3 = \begin{pmatrix} 0 & -1 \\ 1 & -3 \end{pmatrix}\begin{pmatrix} 1 & 3 \\ 0 & 1 \end{pmatrix} = \begin{pmatrix} 0 & -1 \\ 1 & 0 \end{pmatrix} = S.$$

Solving for A we find

$$A = ST^{-3}ST^{-4}ST^2.$$

At each stage there may be more than one power of T that makes $|d| < |c|$ so the process is not unique.

To prove the theorem in general it suffices to consider those matrices $A = \begin{pmatrix} a & b \\ c & d \end{pmatrix}$ in Γ with $c \geq 0$. We use induction on c.

If $c = 0$ then $ad = 1$ so $a = d = \pm 1$ and

$$A = \begin{pmatrix} \pm 1 & b \\ 0 & \pm 1 \end{pmatrix} = \begin{pmatrix} 1 & \pm b \\ 0 & 1 \end{pmatrix} = T^{\pm b}.$$

Thus, A is a power of T.

If $c = 1$ then $ad - b = 1$ so $b = ad - 1$ and

$$A = \begin{pmatrix} a & ad - 1 \\ 1 & d \end{pmatrix} = \begin{pmatrix} 1 & a \\ 0 & 1 \end{pmatrix}\begin{pmatrix} 0 & -1 \\ 1 & 0 \end{pmatrix}\begin{pmatrix} 1 & d \\ 0 & 1 \end{pmatrix} = T^a S T^d.$$

Now assume the theorem has been proved for all matrices A with lower left-hand element $< c$ for some $c \geq 1$. Since $ad - bc = 1$ we have $(c, d) = 1$. Dividing d by c we get

$$d = cq + r, \quad \text{where } 0 < r < c.$$

Then

$$A T^{-q} = \begin{pmatrix} a & b \\ c & d \end{pmatrix}\begin{pmatrix} 1 & -q \\ 0 & 1 \end{pmatrix} = \begin{pmatrix} a & -aq + b \\ c & r \end{pmatrix}$$

and

$$A T^{-q} S = \begin{pmatrix} a & -aq + b \\ c & r \end{pmatrix}\begin{pmatrix} 0 & -1 \\ 1 & 0 \end{pmatrix} = \begin{pmatrix} -aq + b & -a \\ r & -c \end{pmatrix}.$$

By the induction hypothesis, the last matrix is a product of powers of S and T, so A is too. This completes the proof. $\square$

2.3 Fundamental regions

Let G denote any subgroup of the modular group Γ. Two points τ and τ' in the upper half-plane H are said to be *equivalent* under G if $\tau' = A\tau$ for some A in G. This is an equivalence relation since G is a group.

This equivalence relation divides the upper half-plane H into a disjoint collection of equivalence classes called *orbits*. The orbit $G\tau$ is the set of all complex numbers of the form $A\tau$ where $A \in G$.

We select one point from each orbit; the union of all these points is called a *fundamental set* of G. To deal with sets having nice topological properties we modify the concept slightly and define a *fundamental region* of G as follows.

Definition. Let G be a subgroup of the modular group Γ. An open subset R_G of H is called a *fundamental region of G* if it has the following two properties:

(a) No two distinct points of R_G are equivalent under G.
(b) If $\tau \in H$ there is a point τ' in the closure of R_G such that τ' is equivalent to τ under G.

For example, the next theorem will show that a fundamental region R_Γ of the full modular group Γ consists of all τ in H satisfying the inequalities

$$|\tau| > 1, \qquad |\tau + \bar{\tau}| < 1.$$

This region is the shaded portion of Figure 2.1.

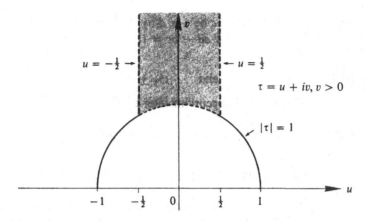

Figure 2.1   Fundamental region of the modular group

The proof will use the following lemma concerning fundamental pairs of periods.

Lemma 1. *Given ω_1', ω_2' with ω_2'/ω_1' not real, let*

$$\Omega = \{m\omega_1' + n\omega_2' : m, n \text{ integers}\}.$$

Then there exists a fundamental pair (ω_1, ω_2) equivalent to (ω_1', ω_2') such that

$$\begin{pmatrix} \omega_2 \\ \omega_1 \end{pmatrix} = \begin{pmatrix} a & b \\ c & d \end{pmatrix}\begin{pmatrix} \omega_2' \\ \omega_1' \end{pmatrix} \quad \text{with } ad - bc = 1,$$

and such that

$$|\omega_2| \geq |\omega_1|, \qquad |\omega_1 + \omega_2| \geq |\omega_2|, \qquad |\omega_1 - \omega_2| \geq |\omega_2|.$$

PROOF. We arrange the elements of Ω in a sequence according to increasing distances from the origin, say

$$\Omega = \{0, w_1, w_2, \ldots\}$$

where

$$0 < |w_1| \leq |w_2| \leq \cdots \quad \text{and} \quad \arg w_n < \arg w_{n+1} \quad \text{if} \quad |w_n| = |w_{n+1}|.$$

31

Let $\omega_1 = w_1$ and let ω_2 be the first member of this sequence that is not a multiple of ω_1. Then the triangle with vertices $0, \omega_1, \omega_2$ contains no element of Ω except the vertices, so (ω_1, ω_2) is a fundamental pair which spans the set Ω. Therefore there exist integers a, b, c, d with $ad - bc = \pm 1$ such that

$$\begin{pmatrix} \omega_2 \\ \omega_1 \end{pmatrix} = \begin{pmatrix} a & b \\ c & d \end{pmatrix} \begin{pmatrix} \omega_2' \\ \omega_1' \end{pmatrix}.$$

If $ad - bc = -1$ we can replace c by $-c$, d by $-d$, and ω_1 by $-\omega_1$ and the same equation holds, except now $ad - bc = 1$. Because of the way we have chosen ω_1, ω_2 we have

$$|\omega_2| \geq |\omega_1| \quad \text{and} \quad |\omega_1 \pm \omega_2| \geq |\omega_2|,$$

since $\omega_1 \pm \omega_2$ are periods in Ω occurring later than ω_2 in the sequence. $\quad\square$

Theorem 2.2. *If $\tau' \in H$, there exists a complex number τ in H equivalent to τ' under Γ such that*

$$|\tau| \geq 1, \quad |\tau + 1| \geq |\tau| \quad \text{and} \quad |\tau - 1| \geq |\tau|.$$

PROOF. Let $\omega_1' = 1$, $\omega_2' = \tau'$ and apply Lemma 1 to the set of periods $\Omega = \{m + n\tau' : m, n \text{ integers}\}$. Then there exists a fundamental pair ω_1, ω_2 with $|\omega_2| \geq |\omega_1|$, $|\omega_1 \pm \omega_2| \geq |\omega_2|$. Let $\tau = \omega_2/\omega_1$. Then $\tau = \begin{pmatrix} a & b \\ c & d \end{pmatrix}\tau'$ with $ad - bc = 1$ and

$$|\tau| \geq 1, \quad |\tau \pm 1| \geq |\tau|. \quad\square$$

Note. Those τ in H satisfying $|\tau \pm 1| \geq |\tau|$ are also those satisfying $|\tau + \bar{\tau}| \leq 1$.

Theorem 2.3. *The open set*

$$R_\Gamma = \{\tau \in H : |\tau| > 1, |\tau + \bar{\tau}| < 1\}$$

is a fundamental region for Γ. Moreover, if $A \in \Gamma$ and if $A\tau = \tau$ for some τ in R_Γ, then $A = I$. In other words, only the identity element has fixed points in R_Γ.

PROOF. Theorem 2.2 shows that if $\tau' \in H$ there is a point τ in the closure of R_Γ equivalent to τ' under Γ. To prove that no two distinct points of R_Γ are equivalent under Γ, let $\tau' = A\tau$ where $A = \begin{pmatrix} a & b \\ c & d \end{pmatrix}$. We show first that $\text{Im}(\tau') < \text{Im}(\tau)$ if $\tau \in R_\Gamma$ and $c \neq 0$. We have

$$\text{Im}(\tau') = \frac{\text{Im}(\tau)}{|c\tau + d|^2}.$$

If $\tau \in R_\Gamma$ and $c \neq 0$ we have

$$|c\tau + d|^2 = (c\tau + d)(c\bar{\tau} + d) = c^2\tau\bar{\tau} + cd(\tau + \bar{\tau}) + d^2 > c^2 - |cd| + d^2.$$

If $d = 0$ we find $|c\tau + d|^2 > c^2 \geq 1$. If $d \neq 0$ we have

$$c^2 - |cd| + d^2 = (|c| - |d|)^2 + |cd| \geq |cd| \geq 1$$

so again $|c\tau + d|^2 > 1$. Therefore $c \neq 0$ implies $|c\tau + d|^2 > 1$ and hence $\text{Im}(\tau') < \text{Im}(\tau)$. In other words, every element A of Γ with $c \neq 0$ decreases the ordinate of each point τ in R_Γ.

Now suppose both τ and τ' are equivalent interior points of R_Γ. Then

$$\tau' = \frac{a\tau + b}{c\tau + d} \qquad \text{and} \qquad \tau = \frac{d\tau' - b}{-c\tau' + a}.$$

If $c \neq 0$ we have both $\text{Im}(\tau') < \text{Im}(\tau)$ and $\text{Im}(\tau) < \text{Im}(\tau')$. Therefore $c = 0$ so $ad = 1$, $a = d = \pm 1$, and

$$A = \begin{pmatrix} a & b \\ c & d \end{pmatrix} = \begin{pmatrix} \pm 1 & b \\ 0 & \pm 1 \end{pmatrix} = T^{\pm b}.$$

But then $b = 0$ since both τ and τ' are in R_Γ so $\tau = \tau'$. This proves that no two distinct points of R_Γ are equivalent under Γ.

Finally, if $A\tau = \tau$ for some τ in R_Γ, the same argument shows that $c = 0$, $a = d = \pm 1$, so $A = I$. This proves that only the identity element has fixed points in R_Γ. $\qquad\square$

Figure 2.2 shows the fundamental region R_Γ and some of its images under transformations of the modular group. Each element of Γ maps circles into circles (where, as usual, straight lines are considered as special cases of circles). Since the boundary curves of R_Γ are circles orthogonal to the real

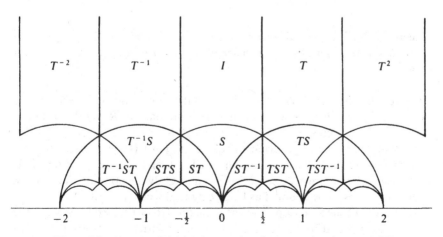

Figure 2.2   Images of the fundamental region R_Γ under elements of Γ

axis, the same is true of every image $f(R_\Gamma)$ under the elements f of Γ. The set of all images $f(R_\Gamma)$, where $f \in \Gamma$, is a collection of nonoverlapping open regions which, together with their boundary points, cover all of H.

2.4 Modular functions

Definition. A function f is said to be *modular* if it satisfies the following three conditions:

(a) f is meromorphic in the upper half-plane H.
(b) $f(A\tau) = f(\tau)$ for every A in the modular group Γ.
(c) The Fourier expansion of f has the form

$$f(\tau) = \sum_{n=-m}^{\infty} a(n)e^{2\pi i n\tau}.$$

Property (a) states that f is analytic in H except possibly for poles. Property (b) states that f is invariant under all transformations of Γ. Property (c) is a condition on the behavior of f at the point $\tau = i\infty$. If $x = e^{2\pi i\tau}$ the Fourier series in (c) is a Laurent expansion in powers of x. The behavior of f at $i\infty$ is described by the nature of this Laurent expansion near 0. If $m > 0$ and $a(-m) \neq 0$ we say that f has a pole of order m at $i\infty$. If $m \leq 0$ we say f is analytic at $i\infty$. Condition (c) states that f has at worst a pole of order m at $i\infty$.

The function J is a modular function. It is analytic in H with a first order pole at $i\infty$. Later we show that every modular function can be expressed as a rational function of J. The proof of this depends on the following property of modular functions.

Theorem 2.4. *If f is modular and not identically zero, then in the closure of the fundamental region R_Γ, the number of zeros of f is equal to the number of poles.*

Note. This theorem is valid only with suitable conventions at the boundary points of R_Γ. First of all, we consider the boundary of R_Γ as the union of four edges intersecting at four vertices $\rho, i, \rho + 1$, and $i\infty$, where $\rho = e^{2\pi i/3}$ (see Figure 2.3). The edges occur in equivalent pairs (1), (4) and (2), (3).

If f has a zero or pole at a point on an edge, then it also has a zero or pole at the equivalent point on the equivalent edge. Only the point on the leftmost edge (1) or (2) is to be counted as belonging to the closure of R_Γ.

The order of the zero or pole at the vertex ρ is to be divided by 3; the order at i is to be divided by 2; the order at $i\infty$ is the order of the zero or pole at $x = 0$, measured in the variable $x = e^{2\pi i\tau}$.

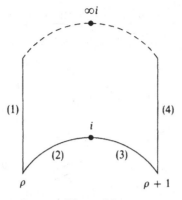

Figure 2.3

PROOF. Assume first that f has no zeros or poles on the finite part of the boundary of R_Γ. Cut R_Γ by a horizontal line, $\mathrm{Im}(\tau) = M$, where $M > 0$ is taken so large that all the zeros or poles of f are inside the truncated region which we call R. [If f had an infinite number of poles in R_Γ they would have an accumulation point at $i\infty$, contradicting condition (c). Similarly, since f is not identically zero, f cannot have an infinite number of zeros in R_Γ.] Let ∂R denote the boundary of the truncated region R. (See Figure 2.4.)

Let N and P denote the number of zeros and poles of f inside R. Then

$$N - P = \frac{1}{2\pi i} \int_{\partial R} \frac{f'(\tau)}{f(\tau)} \, d\tau = \frac{1}{2\pi i} \left\{ \int_{(1)} + \int_{(2)} + \int_{(3)} + \int_{(4)} + \int_{(5)} \right\}$$

where the path is split into five parts as indicated in Figure 2.5. The integrals along (1) and (4) cancel because of periodicity. They also cancel along (2) and (3) because (2) gets mapped onto (3) with a reversal of direction under

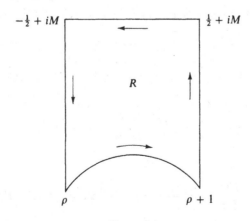

Figure 2.4

35

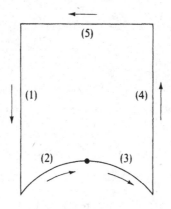

Figure 2.5

the mapping $u = S(\tau) = -1/\tau$, or $\tau = S^{-1}u = S(u)$. The integrand remains unchanged because $f[S(u)] = f(u)$ implies $f'[S(u)]S'(u) = f'(u)$ so

$$\frac{f'(\tau)}{f(\tau)} d\tau = \frac{f'[S(u)]}{f[S(u)]} S'(u) du = \frac{f'(u)}{f(u)} du.$$

Thus we are left with

$$N - P = \frac{1}{2\pi i} \int_{(5)} \frac{f'(\tau)}{f(\tau)} d\tau.$$

We transform this integral to the x-plane, $x = e^{2\pi i \tau}$. As τ varies on the horizontal segment $\tau = u + iM$, $-\frac{1}{2} \le u \le \frac{1}{2}$, we have

$$x = e^{2\pi i(u + iM)} = e^{-2\pi M} e^{2\pi i u}$$

so x varies once around a circle K of radius $e^{-2\pi M}$ about $x = 0$ in the negative direction. The points above this segment are mapped inside K, so f has no zeros or poles inside K, except possibly at $x = 0$. The Fourier expansion gives us

$$f(\tau) = \frac{a_{-m}}{x^m} + \cdots = F(x),$$

say, with

$$f'(\tau) = F'(x)\frac{dx}{d\tau}, \qquad \frac{f'(\tau)}{f(\tau)} d\tau = \frac{F'(x)}{F(x)} dx.$$

Hence

$$N - P = \frac{1}{2\pi i} \int_{(5)} \frac{f'(\tau)}{f(\tau)} d\tau = -\frac{1}{2\pi i} \oint_K \frac{F'(x)}{F(x)} dx = -(N_F - P_F) = P_F - N_F,$$

where N_F and P_F are the number of zeros and poles of F inside K.

If there is a pole of order m at $x = 0$ then $m > 0$, $N_F = 0$, $P_F = m$ so $P_F - N_F = m$, and

$$N = P + m.$$

Therefore f takes on the value 0 in R_Γ as often as it takes the value ∞.

If there is a zero of order n at $x = 0$, then $m = -n$ so $P_F = 0$, $N_F = n$, hence

$$N + n = P.$$

Again, f takes the value 0 in R_Γ as often as it takes the value ∞. This proves the theorem if f has no zeros or poles on the finite part of the boundary of R_Γ.

If f has a zero or a pole on an edge but not at a vertex, we introduce detours in the path of integration so as to include the zero or pole in the interior of R, as indicated in Figure 2.6. The integrals along equivalent edges cancel as before. Only one member of each pair of new zeros or poles lies inside the new region and the proof goes through as before, since by our convention only one of the equivalent points (zero or pole) is considered as belonging to the closure of R_Γ.

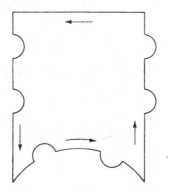

Figure 2.6

If f has a zero or pole at a vertex ρ or i we further modify the path of integration with new detours as indicated in Figure 2.7. Arguing as above we find

$$N - P = \frac{1}{2\pi i}\left\{\left(\int_{C_1} + \int_{C_3}\right) + \int_{C_2} + \int_{1/2+iM}^{-1/2+iM}\right\}\frac{f'(\tau)}{f(\tau)}\,d\tau$$

$$= \frac{1}{2\pi i}\left\{\left(\int_{C_1} + \int_{C_3}\right) + \int_{C_2}\right\}\frac{f'(\tau)}{f(\tau)}\,d\tau + m,$$

where x^{-m} is the lowest power of x occurring in the Laurent expansion near $x = 0$, $x = e^{2\pi i\tau}$.

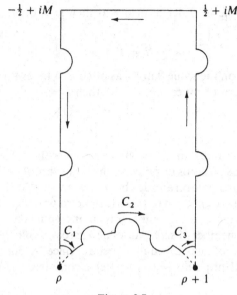

Figure 2.7

Near the vertex ρ we write

$$f(\tau) = (\tau - \rho)^k g(\tau), \quad \text{where } g(\rho) \neq 0.$$

The exponent k is positive if f has a zero at ρ, and negative if f has a pole at ρ. On the path C_1 we write $\tau - \rho = re^{i\theta}$ where r is fixed and $\alpha \leq \theta \leq \pi/2$ where α depends on r. Then

$$\frac{f'(\tau)}{f(\tau)} = \frac{k}{\tau - \rho} + \frac{g'(\tau)}{g(\tau)}$$

and

$$\frac{1}{2\pi i} \int_{C_1} \frac{f'(\tau)}{f(\tau)} \, d\tau = \frac{1}{2\pi i} \int_{\pi/2}^{\alpha} \left(\frac{k}{re^{i\theta}} + \frac{g'(\rho + re^{i\theta})}{g(\rho + re^{i\theta})} \right) re^{i\theta} i \, d\theta$$

$$= \frac{-k\alpha'}{2\pi} + \frac{r}{2\pi} \int_{\pi/2}^{\alpha} \frac{g'(\rho + re^{i\theta})}{g(\rho + re^{i\theta})} e^{i\theta} \, d\theta, \quad \text{where } \alpha' = \frac{\pi}{2} - \alpha.$$

As $r \to 0$, the last term tends to 0 since the integrand is bounded. Also, $\alpha' \to \pi/3$ as $r \to 0$ so

$$\lim_{r \to 0} \frac{1}{2\pi i} \int_{C_1} \frac{f'(\tau)}{f(\tau)} \, d\tau = -\frac{k}{6}.$$

Similarly,

$$\lim_{r \to 0} \frac{1}{2\pi i} \int_{C_3} \frac{f'(\tau)}{f(\tau)} \, d\tau = -\frac{k}{6}$$

so

$$\lim_{r \to 0} \frac{1}{2\pi i} \left(\int_{C_1} + \int_{C_3} \right) \frac{f'(\tau)}{f(\tau)} \, d\tau = -\frac{k}{3}.$$

Similarly, near the vertex i we write

$$f(\tau) = (\tau - i)^l h(\tau), \quad \text{where } h(i) \neq 0$$

and we find, in the same way,

$$\lim_{r \to 0} \frac{1}{2\pi i} \int_{C_2} \frac{f'(\tau)}{f(\tau)} \, d\tau = -\frac{l}{2}.$$

Therefore we get the formula

$$N - P = m - \frac{k}{3} - \frac{l}{2}.$$

If f has a pole at $x = 0$, and zeros at ρ and i, then m, k and l are positive and we have

$$N + \frac{k}{3} + \frac{l}{2} = P + m.$$

The left member counts the number of zeros of f in the closure of R_Γ (with the conventions agreed on at the vertices) and the right member counts the number of poles. If f has a zero of order n at $x = 0$ then $m = -n$ and the equation becomes

$$N + n + \frac{k}{3} + \frac{l}{2} = P.$$

Similarly, if f has a pole at ρ or at i the corresponding term $k/3$ or $l/2$ is negative and gets counted along with P. This completes the proof. $\square$

Theorem 2.5. *If f is modular and not constant, then for every complex c the function $f - c$ has the same number of zeros as poles in the closure of R_Γ. In other words, f takes on every value equally often in the closure of R_Γ.*

PROOF. Apply the previous theorem to $f - c$. $\square$

Theorem 2.6. *If f is modular and bounded in H then f is constant.*

PROOF. Since f is bounded it omits a value so f is constant. $\square$

2.5 Special values of J

Theorem 2.7. *The function J takes every value exactly once in the closure of R_Γ. In particular, at the vertices we have*

$$J(\rho) = 0, \qquad J(i) = 1, \qquad J(i\infty) = \infty.$$

There is a first order pole at $i\infty$, a triple zero at ρ, and $J(\tau) - 1$ has a double zero at $\tau = i$.

PROOF. First we verify that $g_2(\rho) = 0$ and $g_3(i) = 0$. Since $\rho^3 = 1$ and $\rho^2 + \rho + 1 = 0$ we have

$$\frac{1}{60} g_2(\rho) = \sum_{m,n} \frac{1}{(m+n\rho)^4} = \sum_{m,n} \frac{1}{(m\rho^3 + n\rho)^4} = \frac{1}{\rho^4} \sum_{m,n} \frac{1}{(m\rho^2 + n)^4}$$

$$= \frac{1}{\rho} \sum_{m,n} \frac{1}{(n - m - m\rho)^4} = \frac{1}{\rho} \sum_{M,N} \frac{1}{(N + M\rho)^4} = \frac{1}{60\rho} g_2(\rho),$$

so $g_2(\rho) = 0$. A similar argument shows that $g_3(i) = 0$. Therefore

$$J(\rho) = \frac{g_2{}^3(\rho)}{\Delta(\rho)} = 0 \quad \text{and} \quad J(i) = \frac{g_2{}^3(i)}{g_2{}^3(i)} = 1.$$

The multiplicities are a consequence of Theorem 2.4. $\qquad\square$

2.6 Modular functions as rational functions of J

Theorem 2.8. *Every rational function of J is a modular function. Conversely, every modular function can be expressed as a rational function of J.*

PROOF. The first part is clear. To prove the second, suppose f has zeros at $z_1, \ldots, z_n$ and poles at $p_1, \ldots, p_n$ with the usual conventions about multiplicities. Let

$$g(\tau) = \prod_{k=1}^{n} \frac{J(\tau) - J(z_k)}{J(\tau) - J(p_k)}$$

where a factor 1 is inserted whenever z_k or p_k is ∞. Then g has the same zeros and poles as f in the closure of R_Γ, each with proper multiplicity. Therefore f/g has no zeros or poles and must be constant, so f is a rational function of J. $\qquad\square$

2.7 Mapping properties of J

Theorem 2.7 shows that J takes every value exactly once in the closure of the fundamental region R_Γ. Figure 2.8 illustrates how R_Γ is mapped by J onto the complex plane.

The left half of R_Γ (the shaded portion of Figure 2.8a) is mapped onto the upper half-plane (shaded in Figure 2.8b) with the vertical part of the boundary mapping onto the real interval $(-\infty, 0]$. The circular part of the boundary maps onto the interval $[0, 1]$, and the portion of the imaginary axis $v > 1$, $u = 0$ maps onto the interval $(1, +\infty)$. Points in R_Γ symmetric about the imaginary axis map onto conjugate points in $J(R_\Gamma)$. The mapping is conformal except at the vertices $\tau = i$ and $\tau = \rho$ where angles are doubled and tripled, respectively.

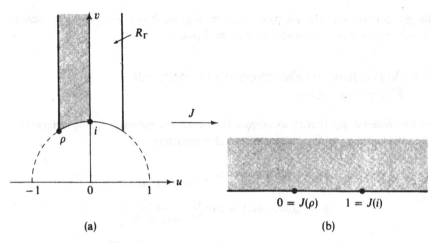

Figure 2.8

These mapping properties can be demonstrated as follows. On the imaginary axis in R_Γ we have $\tau = iv$ hence $x = e^{2\pi i\tau} = e^{-2\pi v} > 0$, so the Fourier series

$$12^3 J(\tau) = \frac{1}{x} + \sum_{n=0}^{\infty} c(n)x^n \quad (x = e^{2\pi i\tau})$$

shows that $J(iv)$ is real. Since $J(i) = 1$ and $J(iv) \to +\infty$ as $v \to +\infty$ the portion of the imaginary axis $1 \le v < +\infty$ gets mapped onto the real axis $1 \le J(\tau) < +\infty$.

On the left boundary of R_Γ we have $\tau = -\frac{1}{2} + iv$, hence $x = e^{2\pi i\tau} = e^{-2\pi v}e^{-\pi i} = -e^{-2\pi v} < 0$. For large v (small x) we have $J(-\frac{1}{2} + iv) < 0$ so J maps the line $u = -\frac{1}{2}$ onto the negative real axis. Since $J(\rho) = 0$ and $J(\infty) = \infty$, the left boundary of R_Γ is mapped onto the line $-\infty < J(\tau) \le 0$. As the boundary of R_Γ is traversed counterclockwise the points inside R_Γ lie on the left, hence the image points lie above the real axis in the image plane.

Finally, we show that J takes conjugate values at points symmetric about the imaginary axis, that is,

$$J(\tau) = \overline{J(-\bar{\tau})}.$$

To see this, write $\tau = u + iv$. Then

$$x = e^{2\pi i\tau} = e^{2\pi i(u + iv)} = e^{-2\pi v}e^{2\pi iu}$$

and

$$\bar{x} = e^{-2\pi v}e^{-2\pi iu} = e^{2\pi i(-u + iv)} = e^{-2\pi i\bar{\tau}}.$$

Thus τ and $-\bar{\tau}$ correspond to conjugate points x and $\bar{x}$, but the Fourier series for J has real coefficients so $J(\tau)$ and $J(-\bar{\tau})$ are complex conjugates.

41

In particular, on the circular arc $\tau\bar{\tau} = 1$ we have $-\bar{\tau} = -1/\tau$, hence $J(-\bar{\tau}) = J(-1/\tau) = J(\tau)$ so J is real on this arc.

2.8 Application to the inversion problem for Eisenstein series

In the Weierstrass theory of elliptic functions the periods ω_1, ω_2 determine the invariants g_2 and g_3 according to the equations

(4)
$$g_2 = g_2(\omega_1, \omega_2) = 60 \sum \frac{1}{(m\omega_1 + n\omega_2)^4}$$
$$g_3 = g_3(\omega_1, \omega_2) = 140 \sum \frac{1}{(m\omega_1 + n\omega_2)^6}.$$

A fundamental problem is to decide whether or not the invariants g_2 and g_3 can take arbitrary prescribed values, subject only to the necessary condition $g_2{}^3 - 27g_3{}^2 \neq 0$. This is called the *inversion problem* for Eisenstein series since it amounts to solving the equations in (4) for ω_1 and ω_2 in terms of g_2 and g_3. The next theorem shows that the problem has a solution.

Theorem 2.9. *Given two complex numbers a_2 and a_3 such that $a_2{}^3 - 27a_3{}^2 \neq 0$. Then there exist complex numbers ω_1 and ω_2 whose ratio is not real such that*

$$g_2(\omega_1, \omega_2) = a_2 \quad and \quad g_3(\omega_1, \omega_2) = a_3.$$

PROOF. We consider three cases: (1) $a_2 = 0$; (2) $a_3 = 0$; (3) $a_2 a_3 \neq 0$.

Case 1. If $a_2 = 0$ then $a_3 \neq 0$ since $a_2{}^3 - 27a_3{}^2 \neq 0$. Let ω_1 be any complex number such that

$$\omega_1{}^6 = \frac{g_3(1, \rho)}{a_3}$$

and let $\omega_2 = \rho\omega_1$, where $\rho = e^{2\pi i/3}$. We know that $g_3(1, \rho) \neq 0$ because $g_2(1, \rho) = 0$ and $\Delta(1, \rho) = g_2{}^3 - 27g_3{}^2 \neq 0$. Then

$$g_2(\omega_1, \omega_2) = g_2(\omega_1, \omega_1\rho) = \frac{1}{\omega_1{}^4} g_2(1, \rho) = 0 = a_2$$

and

$$g_3(\omega_1, \omega_2) = g_3(\omega_1, \omega_1\rho) = \frac{1}{\omega_1{}^6} g_3(1, \rho) = a_3.$$

Case 2. If $a_3 = 0$ then $a_2 \neq 0$ and we take ω_1 to satisfy

$$\omega_1{}^4 = \frac{g_2(1, i)}{a_2}$$

and let $\omega_2 = i\omega_1$. Then

$$g_2(\omega_1, \omega_2) = g_2(\omega_1, i\omega_1) = \frac{1}{\omega_1^4} g_2(1, i) = a_2$$

and

$$g_3(\omega_1, \omega_2) = g_3(\omega_1, i\omega_1) = \frac{1}{\omega_1^6} g_3(1, i) = 0 = a_3.$$

Case 3. Assume $a_2 \neq 0$ and $a_3 \neq 0$. Choose a complex τ with Im $\tau > 0$ such that

$$J(\tau) = \frac{a_2^3}{a_2^3 - 27a_3^2}.$$

Note that $J(\tau) \neq 0$ since $a_2 \neq 0$ and that

(5)
$$\frac{J(\tau) - 1}{J(\tau)} = \frac{27a_3^2}{a_2^3}.$$

For this τ choose ω_1 to satisfy

$$\omega_1^2 = \frac{a_2}{a_3} \cdot \frac{g_3(1, \tau)}{g_2(1, \tau)}$$

and let $\omega_2 = \tau\omega_1$. Then

$$\frac{g_2(\omega_1, \omega_2)}{g_3(\omega_1, \omega_2)} = \frac{\omega_1^{-4} g_2(1, \tau)}{\omega_1^{-6} g_3(1, \tau)} = \omega_1^2 \frac{g_2(1, \tau)}{g_3(1, \tau)} = \frac{a_2}{a_3},$$

so

(6)
$$g_3(\omega_1, \omega_2) = \frac{a_3}{a_2} g_2(\omega_1, \omega_2).$$

But we also have

$$\frac{J(\tau) - 1}{J(\tau)} = \frac{27g_3^2(\omega_1, \omega_2)}{g_2^3(\omega_1, \omega_2)} = \frac{27(a_3/a_2)^2 g_2^2(\omega_1, \omega_2)}{g_2^3(\omega_1, \omega_2)} = \frac{27a_3^2}{a_2^2 g_2(\omega_1, \omega_2)}.$$

Comparing this with (5) we find that $g_2(\omega_1, \omega_2) = a_2$ and hence by (6) we also have $g_3(\omega_1, \omega_2) = a_3$. This completes the proof. $\square$

2.9 Application to Picard's theorem

The modular function J can be used to give a short proof of a famous theorem of Picard in complex analysis.

Theorem 2.10. *Every nonconstant entire function attains every complex value with at most one exception.*

Note. An example is the exponential function $f(z) = e^z$ which omits only the value 0.

PROOF. We assume f is an entire function which omits two values, say a and b, $a \neq b$, and show that f is constant. Let

$$g(z) = \frac{f(z) - a}{b - a}.$$

Then g is entire and omits the values 0 and 1.

The upper half-plane H is covered by the images of the closure of the fundamental region R_Γ under transformations of Γ. Since J maps the closure of R_Γ onto the complex plane, J maps the half-plane H onto an infinite-sheeted Riemann surface with branch points over the points 0, 1 and ∞ (the images of the vertices ρ, i and ∞, respectively). The inverse function J^{-1} maps the Riemann surface back onto the closure of the fundamental region R_Γ. Since $J'(\tau) \neq 0$ if $\tau \neq \rho$ or $\tau \neq i$ and since $J'(\rho) = J'(i) = 0$, each single-valued branch of J^{-1} is locally analytic everywhere except at $0 = J(\rho)$, $1 = J(i)$, and $\infty = J(\infty)$. For each single-valued branch of J^{-1} the composite function

$$h(z) = J^{-1}[g(z)]$$

is a single-valued function element which is locally analytic at each finite z since $g(z)$ is never 0 or 1. Therefore h is arbitrarily continuable in the entire finite z-plane. By the monodromy theorem, the continuation of h exists as a single-valued function analytic in the entire finite z-plane. Thus h is an entire function and so too is

$$\varphi(z) = e^{ih(z)}.$$

But Im $h(z) > 0$ since $h(z) \in H$ so

$$|\varphi(z)| = e^{-\operatorname{Im} h(z)} < 1.$$

Therefore φ is a bounded entire function which, by Liouville's theorem, must be constant. But this implies h is constant and hence g is constant since $g(z) = J[h(z)]$. Therefore f is constant since $f(z) = a + (b - a)g(z)$. $\square$

Exercises for Chapter 2

In these exercises, Γ denotes the modular group, S and T denote its generators, $S(\tau) = -1/\tau$, $T(\tau) = \tau + 1$, and I denotes the identity element.

1. Find all elements A of Γ which (a) commute with S; (b) commute with ST.

2. Find the smallest integer $n > 0$ such that $(ST)^n = I$.

3. Determine the point τ in the fundamental region R_Γ which is equivalent to
   (a) $(8 + 6i)/(3 + 2i)$; (b) $(10i + 11)/(6i + 12)$.

4. Determine all elements A of Γ which leave i fixed.

5. Determine all elements A of Γ which leave $\rho = e^{2\pi i/3}$ fixed.

QUADRATIC FORMS AND THE MODULAR GROUP

The following exercises relate quadratic forms and the modular group Γ. We consider quadratic forms $Q(x, y) = ax^2 + bxy + cy^2$ in x and y with real coefficients a, b, c. The number $d = 4ac - b^2$ is called the *discriminant* of $Q(x, y)$.

6. If x and y are subjected to a unimodular transformation, say

(1) $x = \alpha x' + \beta y', \quad y = \gamma x' + \delta y', \quad \text{where} \begin{pmatrix} \alpha & \beta \\ \gamma & \delta \end{pmatrix} \in \Gamma,$

prove that $Q(x, y)$ gets transformed to a quadratic form $Q_1(x', y')$ having the same discriminant. Two forms $Q(x, y)$ and $Q_1(x', y')$ so related are called *equivalent*. This equivalence relation separates all forms into equivalence classes. The forms in a given class have the same discriminant, and they represent the same integers. That is, if $Q(x, y) = n$ for some pair of integers x and y, then $Q_1(x', y') = n$ for the pair of integers x', y' given by (1).

In Exercises 7 thru 10 we consider forms $ax^2 + bxy + cy^2$ with $d > 0$, $a > 0$, and $c > 0$. The associated quadratic polynomial

$$f(z) = az^2 + bz + c$$

has two complex roots. The root τ with positive imaginary part is called the *representative* of the quadratic form $Q(x, y) = ax^2 + bxy + cy^2$.

7. (a) If d is fixed, prove that there is a one-to-one correspondence between the set of forms with discriminant d and the set of complex numbers τ with $\text{Im}(\tau) > 0$.
   (b) Prove that two quadratic forms with discriminant d are equivalent if and only if their representatives are equivalent under Γ.

Note. A *reduced form* is one whose representative $\tau \in R_\Gamma$. Thus, two reduced forms are equivalent if and only if they are identical. Also, each class of equivalent forms contains exactly one reduced form.

8. Prove that a form $Q(x, y) = ax^2 + bxy + cy^2$ is reduced if, and only if, either $-a < b \leq a < c$ or $0 \leq b \leq a = c$.

9. Assume now that the form $Q(x, y) = ax^2 + bxy + cy^2$ has integer coefficients a, b, c. Prove that for a given d there are only a finite number of equivalence classes with discriminant d. This number is called the *class number* and is denoted by $h(d)$.

   Hint: Show that $0 < a \leq \sqrt{d/3}$ for each reduced form.

10. Determine all reduced forms with integer coefficients a, b, c and the class number $h(d)$ for each d in the interval $1 \leq d \leq 20$.

CONGRUENCE SUBGROUPS

The modular group Γ has many subgroups of special interest in number theory. The following exercises deal with a class of subgroups called *congruence* subgroups. Let

$$A = \begin{pmatrix} a & b \\ c & d \end{pmatrix} \quad \text{and} \quad B = \begin{pmatrix} \alpha & \beta \\ \gamma & \delta \end{pmatrix}$$

be two unimodular matrices. (In this discussion we do not identify a matrix with its negative.) If n is a positive integer write

$$A \equiv B \pmod{n} \quad \text{whenever } a \equiv \alpha, b \equiv \beta, c \equiv \gamma \text{ and } d \equiv \delta \pmod{n}.$$

This defines an equivalence relation with the property that

$$A_1 \equiv A_2 \pmod{n} \quad \text{and} \quad B_1 \equiv B_2 \pmod{n}$$

implies

$$A_1 B_1 \equiv A_2 B_2 \pmod{n} \quad \text{and} \quad A_1^{-1} \equiv A_2^{-1} \pmod{n}.$$

Hence

$$A \equiv B \pmod{n} \quad \text{if, and only if, } AB^{-1} \equiv I \pmod{n},$$

where I is the identity matrix. We denote by $\Gamma^{(n)}$ the set of all matrices in Γ congruent modulo n to the identity. This is called the *congruence subgroup of level n* (*stufe n*, in German).

Prove each of the following statements:

11. $\Gamma^{(n)}$ is a subgroup of Γ. Moreover, if $B \in \Gamma^{(n)}$ then $A^{-1}BA \in \Gamma^{(n)}$ for every A in Γ. That is, $\Gamma^{(n)}$ is a normal subgroup of Γ.

12. The quotient group $\Gamma / \Gamma^{(n)}$ is finite. That is, there exist a finite number of elements of Γ, say $A_1, \ldots, A_k$, such that every B in Γ is representable in the form

$$B = A_i B^{(n)} \quad \text{where } 1 \leq i \leq k \quad \text{and } B^{(n)} \in \Gamma^{(n)}.$$

The smallest such k is called the index of $\Gamma^{(n)}$ in Γ.

13. The index of $\Gamma^{(n)}$ in Γ is the number of equivalence classes of matrices modulo n.

The following exercises determine an explicit formula for the index.

14. Given integers a, b, c, d with $ad - bc \equiv 1 \pmod{n}$, there exist integers $\alpha, \beta, \gamma, \delta$ such that $\alpha \equiv a, \beta \equiv b, \gamma \equiv c, \delta \equiv d \pmod{n}$ with $\alpha\delta - \beta\gamma = 1$.

15. If $(m, n) = 1$ and $A \in \Gamma$ there exists $\bar{A}$ in Γ such that

$$\bar{A} \equiv A \pmod{n} \quad \text{and} \quad \bar{A} \equiv I \pmod{m}.$$

16. Let $f(n)$ denote the number of equivalence classes of matrices modulo n. Then f is a multiplicative function.

17. If a, b, n are integers with $n \geq 1$ and $(a, b, n) = 1$ the congruence

$$ax - by \equiv 1 \pmod{n}$$

has exactly n solutions, distinct mod n.(A solution is an ordered pair (x, y) of integers.)

18. For each prime p the number of solutions, distinct mod p^r, of all possible congruences of the form

$$ax - by \equiv 1 \pmod{p^r}, \quad \text{where } (a, b, p) = 1,$$

is equal to $f(p^r)$.

19. If p is prime the number of pairs of integers (a, b), incongruent mod p^r, which satisfy the condition $(a, b, p) = 1$ is $p^{2r-2}(p^2 - 1)$.

20. $f(n) = n^3 \sum_{d \mid n} \mu(d)/d^2$, where μ is the Möbius function.

The Dedekind eta function · 3

3.1 Introduction

In many applications of elliptic modular functions to number theory the eta function plays a central role. It was introduced by Dedekind in 1877 and is defined in the half-plane $H = \{\tau : \text{Im}(\tau) > 0\}$ by the equation

$$(1) \qquad \eta(\tau) = e^{\pi i \tau/12} \prod_{n=1}^{\infty} (1 - e^{2\pi i n \tau}).$$

The infinite product has the form $\prod (1 - x^n)$ where $x = e^{2\pi i \tau}$. If $\tau \in H$ then $|x| < 1$ so the product converges absolutely and is nonzero. Moreover, since the convergence is uniform on compact subsets of H, $\eta(\tau)$ is analytic on H.

The eta function is closely related to the discriminant $\Delta(\tau)$ introduced in Chapter 1. Later in this chapter we show that

$$\Delta(\tau) = (2\pi)^{12} \eta^{24}(\tau).$$

This result and other properties of $\eta(\tau)$ follow from transformation formulas which describe the behavior of $\eta(\tau)$ under elements of the modular group Γ. For the generator $T\tau = \tau + 1$ we have

$$(2) \qquad \eta(\tau + 1) = e^{\pi i (\tau + 1)/12} \prod_{n=1}^{\infty} (1 - e^{2\pi i n (\tau + 1)}) = e^{\pi i/12} \eta(\tau).$$

Consequently, for any integer b we have

$$(3) \qquad \eta(\tau + b) = e^{\pi i b/12} \eta(\tau).$$

Equation (2) also shows that $\eta^{24}(\tau)$ is periodic with period 1.

For the other generator $S\tau = -1/\tau$ we have the following theorem.

Theorem 3.1. *If $\tau \in H$ we have*

(4)
$$\eta\left(\frac{-1}{\tau}\right) = (-i\tau)^{1/2}\eta(\tau).$$

Note. We choose that branch of the square root function $z^{1/2}$ which is positive when $z > 0$.

This chapter gives two different proofs of (4). The first is a short proof of C. L. Siegel [48] based on residue calculus, and the second derives (4) as a special case of a more general functional equation which relates

$$\eta\left(\frac{a\tau + b}{c\tau + d}\right)$$

to $\eta(\tau)$ when

$$\begin{pmatrix} a & b \\ c & d \end{pmatrix} \in \Gamma \quad \text{and } c > 0.$$

(See Theorem 3.4.) A third proof, based on interchange of summation in a conditionally convergent iterated series, is outlined in the exercises.

3.2 Siegel's proof of Theorem 3.1

First we prove (4) for $\tau = iy$, where $y > 0$, and then extend the result to all τ in H by analytic continuation. If $\tau = iy$ the transformation formula becomes $\eta(i/y) = y^{1/2}\eta(iy)$, and this is equivalent to

$$\log \eta(i/y) - \log \eta(iy) = \tfrac{1}{2} \log y.$$

Now

$$\log \eta(iy) = -\frac{\pi y}{12} + \log \prod_{n=1}^{\infty} (1 - e^{-2\pi ny})$$

$$= -\frac{\pi y}{12} + \sum_{n=1}^{\infty} \log(1 - e^{-2\pi ny}) = -\frac{\pi y}{12} - \sum_{n=1}^{\infty} \sum_{m=1}^{\infty} \frac{e^{-2\pi mny}}{m}$$

$$= -\frac{\pi y}{12} - \sum_{m=1}^{\infty} \frac{1}{m} \frac{e^{-2\pi my}}{1 - e^{-2\pi my}} = -\frac{\pi y}{12} + \sum_{m=1}^{\infty} \frac{1}{m} \frac{1}{1 - e^{2\pi my}}.$$

Therefore we are to prove that

(5)
$$\sum_{m=1}^{\infty} \frac{1}{m} \frac{1}{1 - e^{2\pi my}} - \sum_{m=1}^{\infty} \frac{1}{m} \frac{1}{1 - e^{2\pi m/y}} - \frac{\pi}{12}\left(y - \frac{1}{y}\right) = -\frac{1}{2} \log y.$$

This will be proved with the help of residue calculus.

For fixed $y > 0$ and $n = 1, 2, \ldots$, let

$$F_n(z) = -\frac{1}{8z} \cot \pi iNz \cot \frac{\pi Nz}{y},$$

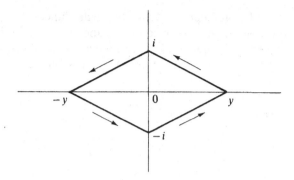

Figure 3.1

where $N = n + \frac{1}{2}$. Let C be the parallelogram joining the vertices $y, i, -y, -i$ in that order. (See Figure 3.1.) Inside C, F_n has simple poles at $z = ik/N$ and at $z = ky/N$ for $k = \pm 1, \pm 2, \ldots, \pm n$. There is also a triple pole at $z = 0$ with residue $i(y - y^{-1})/24$. The residue at $z = ik/N$ is

$$\frac{1}{8\pi k} \cot \frac{\pi i k}{y}.$$

Since this is an even function of k we have

$$\sum_{\substack{k=-n \\ k \neq 0}}^{n} \operatorname*{Res}_{z=ik/N} F_n(z) = 2 \sum_{k=1}^{n} \frac{1}{8\pi k} \cot \frac{\pi i k}{y}.$$

But

$$\cot i\theta = \frac{\cos i\theta}{\sin i\theta} = i \frac{e^{-\theta} + e^{\theta}}{e^{-\theta} - e^{\theta}} = -i \frac{e^{2\theta} + 1}{e^{2\theta} - 1} = \frac{1}{i}\left(1 - \frac{2}{1 - e^{2\theta}}\right).$$

Using this with $\theta = \pi k/y$ we get

$$\sum_{\substack{k=-n \\ k \neq 0}}^{n} \operatorname*{Res}_{z=ik/N} F_n(z) = \frac{1}{4\pi i} \sum_{k=1}^{n} \frac{1}{k} - \frac{1}{2\pi i} \sum_{k=1}^{n} \frac{1}{k} \frac{1}{1 - e^{2\pi k/y}}.$$

Similarly

$$\sum_{\substack{k=-n \\ k \neq 0}}^{n} \operatorname*{Res}_{z=ky/N} F_n(z) = \frac{i}{4\pi} \sum_{k=1}^{n} \frac{1}{k} - \frac{i}{2\pi} \sum_{k=1}^{n} \frac{1}{k} \frac{1}{1 - e^{2\pi k y}}.$$

Hence $2\pi i$ times the sum of all the residues of $F_n(z)$ inside C is an expression whose limit as $n \to \infty$ is equal to the left member of (5). Therefore, to complete the proof we need only show that

$$\lim_{n \to \infty} \int_C F_n(z)\, dz = -\tfrac{1}{2} \log y.$$

49

On the edges of C (except at the vertices) the function $zF_n(z)$ has, as $n \to \infty$, the limit $\frac{1}{8}$ on the edges connecting y, i and $-y$, $-i$, and the limit $-\frac{1}{8}$ on the other two edges. Moreover, $F_n(z)$ is uniformly bounded on C for all n (because $N = n + \frac{1}{2}$ and $y > 0$). Hence by Arzelà's bounded convergence theorem (Theorem 9.12 in [3]) we have

$$\lim_{n \to \infty} \int_C F_n(z)\, dz = \int_C \lim_{n \to \infty} zF_n(z)\, \frac{dz}{z}$$

$$= \frac{1}{8} \left\{ -\int_{-i}^{y} + \int_{y}^{i} - \int_{i}^{-y} + \int_{-y}^{-i} \right\} \frac{dz}{z}$$

$$= \frac{1}{4} \left\{ -\int_{-i}^{y} + \int_{y}^{i} \right\} \frac{dz}{z}$$

$$= \frac{1}{4} \left\{ -\left(\log y + \frac{\pi i}{2} \right) + \left(\frac{\pi i}{2} - \log y \right) \right\} = -\frac{1}{2} \log y.$$

This completes the proof. $\qquad\qquad\qquad\qquad\qquad\qquad\qquad\qquad\qquad\square$

3.3 Infinite product representation for $\Delta(\tau)$

In this section we express the discriminant $\Delta(\tau)$ in terms of $\eta(\tau)$ and thereby obtain a product representation of $\Delta(\tau)$. The result makes use of the following property of $\Delta(\tau)$.

Theorem 3.2. *If* $\begin{pmatrix} a & b \\ c & d \end{pmatrix} \in \Gamma$ *then*

$$\Delta\left(\frac{a\tau + b}{c\tau + d} \right) = (c\tau + d)^{12} \Delta(\tau).$$

In particular,

$$\Delta(\tau + 1) = \Delta(\tau) \qquad and \qquad \Delta\left(-\frac{1}{\tau} \right) = \tau^{12} \Delta(\tau).$$

PROOF. Since $\Delta(\omega_1, \omega_2)$ is homogeneous of degree -12 we have

$$\Delta(\omega_1, \omega_2) = \omega_1^{-12} \Delta(1, \tau) = \omega_1^{-12} \Delta(\tau),$$

where $\tau = \omega_2/\omega_1$. Also,

$$\Delta(\omega_1, \omega_2) = \Delta(\omega_1', \omega_2')$$

if (ω_1, ω_2) and (ω_1', ω_2') are equivalent pairs of periods. Taking $\omega_1 = 1$, $\omega_2 = \tau$, $\omega_1' = c\tau + d$, $\omega_2' = a\tau + b$, we find

$$\Delta(\tau) = \Delta(\omega_1, \omega_2) = \Delta(c\tau + d, a\tau + b) = (c\tau + d)^{-12} \Delta\left(1, \frac{a\tau + b}{c\tau + d} \right). \quad \square$$

Theorem 3.3. *If $\tau \in H$ and $x = e^{2\pi i \tau}$ we have*

(6) $$\Delta(\tau) = (2\pi)^{12}\eta^{24}(\tau) = (2\pi)^{12}x \prod_{n=1}^{\infty}(1 - x^n)^{24}.$$

Consequently,

(7) $$\sum_{n=1}^{\infty}\tau(n)x^n = x \prod_{n=1}^{\infty}(1 - x^n)^{24} \quad \textit{whenever } |x| < 1$$

where $\tau(n)$ is Ramanujan's tau function.

PROOF. Let $f(\tau) = \Delta(\tau)/\eta^{24}(\tau)$. Then $f(\tau + 1) = f(\tau)$ and $f(-1/\tau) = f(\tau)$, so f is invariant under every transformation in Γ. Also, f is analytic and non-zero in H because Δ is analytic and nonzero and η never vanishes in H.

Next we examine the behavior of f at $i\infty$. We have

$$\eta^{24}(\tau) = e^{2\pi i \tau} \prod_{n=1}^{\infty}(1 - e^{2\pi i n \tau})^{24} = x \prod_{n=1}^{\infty}(1 - x^n)^{24} = x(1 + I(x)),$$

where $I(x)$ denotes a power series in x with integer coefficients. Thus, $\eta^{24}(\tau)$ has a first order zero at $x = 0$. By Theorem 1.19 we also have the Fourier expansion

(8) $$\Delta(\tau) = (2\pi)^{12}\sum_{n=1}^{\infty}\tau(n)x^n = (2\pi)^{12}x(1 + I(x)).$$

Thus, near $i\infty$ the function f has the Fourier expansion

(9) $$f(\tau) = \frac{\Delta(\tau)}{\eta^{24}(\tau)} = \frac{(2\pi)^{12}x(1 + I(x))}{x(1 + I(x))} = (2\pi)^{12}(1 + I(x)),$$

so f is analytic and nonzero at $i\infty$. Therefore f is a modular function which never takes the value 0, so f must be constant. Moreover, (9) shows that this constant is $(2\pi)^{12}$, hence $\Delta(\tau) = (2\pi)^{12}\eta^{24}(\tau)$. This proves (6), and (7) follows from (8). $\square$

3.4 The general functional equation for $\eta(\tau)$

Extracting 24th roots in the relation

$$\Delta\left(\frac{a\tau + b}{c\tau + d}\right) = (c\tau + d)^{12}\Delta(\tau)$$

and using (6) we find that

$$\eta\left(\frac{a\tau + b}{c\tau + d}\right) = \varepsilon(c\tau + d)^{1/2}\eta(\tau),$$

where $\varepsilon^{24} = 1$. For many applications of $\eta(\tau)$ we require more explicit information concerning ε. This is provided in the next theorem.

Theorem 3.4 (Dedekind's functional equation). *If* $\begin{pmatrix} a & b \\ c & d \end{pmatrix} \in \Gamma,\ c > 0,$ *and* $\tau \in H,$ *we have*

(10) $$\eta\left(\frac{a\tau + b}{c\tau + d}\right) = \varepsilon(a, b, c, d)\{-i(c\tau + d)\}^{1/2}\eta(\tau)$$

where

$$\varepsilon(a, b, c, d) = \exp\left\{\pi i\left(\frac{a + d}{12c} + s(-d, c)\right)\right\}$$

and

(11) $$s(h, k) = \sum_{r=1}^{k-1} \frac{r}{k}\left(\frac{hr}{k} - \left[\frac{hr}{k}\right] - \frac{1}{2}\right).$$

Note. The sum $s(h, k)$ in (11) is called a *Dedekind sum*. Some of its properties are discussed later in this chapter.

We will prove Theorem 3.4 through a sequence of lemmas. First we note that Dedekind's formula is a consequence of the following equation, obtained by taking logarithms of both members of (10),

(12) $$\log \eta\left(\frac{a\tau + b}{c\tau + d}\right) = \log \eta(\tau) + \pi i\left(\frac{a + d}{12c} + s(-d, c)\right) + \tfrac{1}{2}\log\{-i(c\tau + d)\}.$$

From the definition of $\eta(\tau)$ as a product we have

(13) $$\log \eta(\tau) = \frac{\pi i\tau}{12} + \sum_{n=1}^{\infty} \log(1 - e^{2\pi i n\tau}) = \frac{\pi i\tau}{12} - \sum_{n=1}^{\infty} \lambda(-in\tau),$$

where $\lambda(x)$ is defined for $\mathrm{Re}(x) > 0$ by the equation

(14) $$\lambda(x) = -\log(1 - e^{-2\pi x}) = \sum_{m=1}^{\infty} \frac{e^{-2\pi mx}}{m}.$$

Equations (12) and (13) give us

Lemma 1. *Equation* (12) *is equivalent to the relation*

(15) $$\sum_{n=1}^{\infty} \lambda(-in\tau) = \sum_{n=1}^{\infty} \lambda\left(-in\frac{a\tau + b}{c\tau + d}\right) + \frac{\pi i}{12}\left(\tau - \frac{a\tau + b}{c\tau + d}\right)$$
$$+ \pi i\left(\frac{a + d}{12c} + s(-d, c)\right) + \tfrac{1}{2}\log\{-i(c\tau + d)\}.$$

We shall prove (15) as a consequence of a more general transformation formula obtained by Shô Iseki [17] in 1957. For this purpose it is convenient to restate (15) in an equivalent form which merely involves some changes in notation.

Lemma 2. *Let z be any complex number with* $\mathrm{Re}(z) > 0$, *and let h, k and H be any integers satisfying* $(h, k) = 1, k > 0, hH \equiv -1 \pmod{k}$. *Then Equation (15) is equivalent to the formula*

$$(16) \qquad \sum_{n=1}^{\infty} \lambda\left\{\frac{n}{k}(z - ih)\right\} = \sum_{n=1}^{\infty} \lambda\left\{\frac{n}{k}\left(\frac{1}{z} - iH\right)\right\}$$

$$+ \tfrac{1}{2}\log z - \frac{\pi}{12k}\left(z - \frac{1}{z}\right) + \pi i s(h, k).$$

PROOF. Given $\begin{pmatrix} a & b \\ c & d \end{pmatrix}$ in Γ, with $c > 0$, and given τ with $\mathrm{Im}(\tau) > 0$, choose $z, h, k,$ and H as follows:

$$k = c, \qquad h = -d, \qquad H = a, \qquad z = -i(c\tau + d).$$

Then $\mathrm{Re}(z) > 0$, and the condition $ad - bc = 1$ implies $-hH - bk = 1$, so $(h, k) = 1$ and $hH \equiv -1 \pmod{k}$. Now $b = -(hH + 1)/k$ and $iz = c\tau + d$, so

$$\tau = \frac{iz - d}{c} = \frac{iz + h}{k}$$

and hence

$$a\tau + b = H\frac{iz + h}{k} - \frac{hH + 1}{k} = \frac{iz}{k}\left(H + \frac{i}{z}\right).$$

Therefore, since $c\tau + d = iz$, we have

$$\frac{a\tau + b}{c\tau + d} = \frac{1}{k}\left(H + \frac{i}{z}\right).$$

Consequently

$$\tau - \frac{a\tau + b}{c\tau + d} = \frac{1}{k}(h - H) + \frac{i}{k}\left(z - \frac{1}{z}\right) = -\frac{a + d}{c} + \frac{i}{k}\left(z - \frac{1}{z}\right)$$

so

$$\frac{\pi i}{12}\left(\tau - \frac{a\tau + b}{c\tau + d}\right) = -\pi i\left(\frac{a + d}{12c}\right) - \frac{\pi}{12k}\left(z - \frac{1}{z}\right).$$

Substituting these expressions in (15) we obtain (16). In the same way we find that (16) implies (15). $\qquad\square$

3.5 Iseki's transformation formula

Theorem 3.5 (Iseki's formula). *If* $\mathrm{Re}(z) > 0$ *and* $0 \le \alpha \le 1, 0 \le \beta \le 1$, *let*

$$(17) \qquad \Lambda(\alpha, \beta, z) = \sum_{r=0}^{\infty}\{\lambda((r + \alpha)z - i\beta) + \lambda((r + 1 - \alpha)z + i\beta)\}.$$

53

Then if either $0 \leq \alpha \leq 1$ and $0 < \beta < 1$, or $0 < \alpha < 1$ and $0 \leq \beta \leq 1$, we have

$$(18) \quad \Lambda(\alpha, \beta, z) = \Lambda(1 - \beta, \alpha, z^{-1}) - \pi z \sum_{n=0}^{2} \binom{2}{n}(iz)^{-n} B_{2-n}(\alpha)B_n(\beta).$$

Note. The sum on the right of (18), which contains Bernoulli polynomials $B_n(x)$, is equal to

$$-\pi z\left(\alpha^2 - \alpha + \frac{1}{6}\right) + \frac{\pi}{z}\left(\beta^2 - \beta + \frac{1}{6}\right) + 2\pi i\left(\alpha - \frac{1}{2}\right)\left(\beta - \frac{1}{2}\right).$$

PROOF. First we assume that $0 < \alpha < 1$ and $0 < \beta < 1$. We begin with the first sum appearing in (17) and use (14) to write

$$(19) \qquad \sum_{r=0}^{\infty} \lambda((r + \alpha)z - i\beta) = \sum_{r=0}^{\infty} \sum_{m=1}^{\infty} \frac{e^{2\pi i m\beta}}{m} e^{-2\pi m(r + \alpha)z}.$$

Now we use Mellin's integral for e^{-x} which states that

$$(20) \qquad e^{-x} = \frac{1}{2\pi i} \int_{c-\infty i}^{c+\infty i} \Gamma(s)x^{-s}\, ds,$$

where $c > 0$ and $\text{Re}(x) > 0$. This is a special case of Mellin's inversion formula which states that, under certain regularity conditions, we have

$$\varphi(s) = \int_0^{\infty} x^{s-1}\psi(x)\, dx \quad \text{if, and only if,} \quad \psi(x) = \frac{1}{2\pi i} \int_{c-\infty i}^{c+\infty i} \varphi(s)x^{-s}\, ds.$$

In this case we take $\varphi(s)$ to be the gamma function integral,

$$\Gamma(s) = \int_0^{\infty} x^{s-1}e^{-x}\, dx$$

and invert this to obtain (20). (Mellin's inversion formula can be deduced from the Fourier integral theorem, a proof of which is given in [3]. See also [49], p. 7.) Applying (20) with $x = 2\pi m(r + \alpha)z$ and $c = 3/2$ to the last exponential in (19) and writing $\int_{(c)}$ for $\int_{c-\infty i}^{c+\infty i}$ we obtain

$$\sum_{r=0}^{\infty} \lambda((r + \alpha)z - i\beta) = \sum_{r=0}^{\infty} \sum_{m=1}^{\infty} \frac{e^{2\pi i m\beta}}{m} \frac{1}{2\pi i} \int_{(3/2)} \Gamma(s)\{2\pi m(r + \alpha)z\}^{-s}\, ds$$

$$= \frac{1}{2\pi i} \int_{(3/2)} \frac{\Gamma(s)}{(2\pi z)^s} \sum_{r=0}^{\infty} \frac{1}{(r + \alpha)^s} \sum_{m=1}^{\infty} \frac{e^{2\pi i m\beta}}{m^{1+s}}\, ds$$

$$= \frac{1}{2\pi i} \int_{(3/2)} \frac{\Gamma(s)}{(2\pi z)^s} \zeta(s, \alpha)F(\beta, 1 + s)\, ds.$$

Here $\zeta(s, \alpha)$ is the Hurwitz zeta function and $F(x, s)$ is the periodic zeta function defined, respectively, by the series

$$\zeta(s, \alpha) = \sum_{r=0}^{\infty} \frac{1}{(r + \alpha)^s}, \quad \text{and} \quad F(x, s) = \sum_{m=1}^{\infty} \frac{e^{2\pi i m x}}{m^s}$$

where $\mathrm{Re}(s) > 1$, $0 < \alpha \le 1$, and x is real. In the same way we find

$$\sum_{r=0}^{\infty} \lambda((r + 1 - \alpha)z + i\beta) = \frac{1}{2\pi i} \int_{(3/2)} \frac{\Gamma(s)}{(2\pi z)^s} \zeta(s, 1 - \alpha)F(1 - \beta, 1 + s)\, ds,$$

so (17) becomes

(21) $$\Lambda(\alpha, \beta, z) = \frac{1}{2\pi i} \int_{(3/2)} z^{-s}\Phi(\alpha, \beta, s)\, ds,$$

where

(22) $$\Phi(\alpha, \beta, s) = \frac{\Gamma(s)}{(2\pi)^s} \{\zeta(s, \alpha)F(\beta, 1 + s) + \zeta(s, 1 - \alpha)F(1 - \beta, 1 + s)\}.$$

Now we shift the line of integration from $c = \frac{3}{2}$ to $c = -\frac{3}{2}$. Actually, we apply Cauchy's theorem to the rectangular contour shown in Figure 3.2,

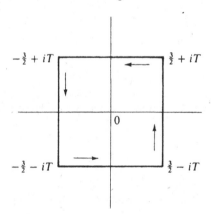

Figure 3.2

and then let $T \to \infty$. In Exercise 8 we show that the integrals along the horizontal segments tend to 0 as $T \to \infty$, so we get

$$\int_{(3/2)} = \int_{(-3/2)} + R$$

where R is the sum of the residues at the poles of the integrand inside the rectangle. This gives us the formula

$$\Lambda(\alpha, \beta, z) = \frac{1}{2\pi i} \int_{(-3/2)} z^{-s}\Phi(\alpha, \beta, s)\, ds + R.$$

In this integral we make the change of variable $u = -s$ to get it back in the form of an integral along the $\frac{3}{2}$ line. This gives us

(23)
$$\Lambda(\alpha, \beta, z) = \frac{1}{2\pi i} \int_{(3/2)} z^u \Phi(\alpha, \beta, -u) \, du + R.$$

Now the function Φ satisfies the functional equation

(24)
$$\Phi(\alpha, \beta, -s) = \Phi(1 - \beta, \alpha, s).$$

This is a consequence of Hurwitz's formula for $\zeta(s, \alpha)$ and a proof is outlined in Exercise 7. Using (24) in (23) we find that

(25)
$$\Lambda(\alpha, \beta, z) = \Lambda(1 - \beta, \alpha, z^{-1}) + R.$$

To complete the proof of Iseki's formula we need to compute the residue sum R.

Equation (22) shows that $\Phi(\alpha, \beta, s)$ has a first order pole at each of the points $s = 1, 0$ and -1. Denoting the corresponding residues by $R(1)$, $R(0)$ and $R(-1)$ we find

$$R(1) = \frac{\Gamma(1)}{2\pi z} \{F(\beta, 2) + F(1 - \beta, 2)\} = \frac{1}{2\pi z} \sum_{n=1}^{\infty} \left(\frac{e^{2\pi i n \beta}}{n^2} + \frac{e^{-2\pi i n \beta}}{n^2} \right)$$

$$= \frac{1}{2\pi z} \sum_{\substack{n=-\infty \\ n \neq 0}}^{\infty} \frac{e^{2\pi i n \beta}}{n^2} = \frac{1}{2\pi z} \frac{-(2\pi i)^2}{2!} B_2(\beta) = \frac{\pi}{z} B_2(\beta),$$

where we have used Theorem 12.19 of [4] to express the Fourier series as a Bernoulli polynomial.

To calculate $R(0)$ we recall that $\zeta(0, \alpha) = \frac{1}{2} - \alpha$. Hence $\zeta(0, 1 - \alpha) = \alpha - \frac{1}{2}$ so

$$R(0) = \zeta(0, \alpha)F(\beta, 1) + \zeta(0, 1 - \alpha)F(1 - \beta, 1) = (\tfrac{1}{2} - \alpha) \sum_{n=1}^{\infty} \frac{e^{2\pi i n \beta} - e^{-2\pi i n \beta}}{n}$$

$$= (\tfrac{1}{2} - \alpha) \sum_{\substack{n=-\infty \\ n \neq 0}}^{\infty} \frac{e^{2\pi i n \beta}}{n} = -B_1(\alpha) \sum_{\substack{n=-\infty \\ n \neq 0}}^{\infty} \frac{e^{2\pi i n \beta}}{n} = 2\pi i B_1(\alpha)B_1(\beta),$$

where again we have used Theorem 12.19 of [4]. To calculate $R(-1)$ we write

$$R(-1) = \operatorname*{Res}_{s=-1} z^{-s}\Phi(\alpha, \beta, s) = \lim_{s \to -1} (s + 1)z^{-s}\Phi(\alpha, \beta, s)$$

$$= \lim_{s \to 1} (-s + 1)z^s\Phi(\alpha, \beta, -s).$$

Using the functional equation (24) we find

$$R(-1) = \lim_{s \to 1}(1 - s)z^s\Phi(1 - \beta, \alpha, s) = -\operatorname*{Res}_{s=1} z^s\Phi(1 - \beta, \alpha, s).$$

Note that this is the same as $R(1) = \text{Res}_{s=1} z^{-s}\Phi(\alpha, \beta, s)$, except that z is replaced by $-z^{-1}$, α by $1 - \beta$, and β by α. Hence we have

$$R(-1) = -\pi z B_2(\alpha).$$

Thus

$$R = R(-1) + R(0) + R(1) = -\pi z \sum_{n=0}^{2} \binom{2}{n}(iz)^{-n}B_{2-n}(\alpha)B_n(\beta).$$

This proves Iseki's formula under the restriction $0 < \alpha < 1, 0 < \beta < 1$.

Finally, we use a limiting argument to show it is valid if $0 \leq \alpha \leq 1$ and $0 < \beta < 1$, or if $0 \leq \beta \leq 1$ and $0 < \alpha < 1$. For example, consider the series

$$\sum_{r=0}^{\infty} \lambda((r + \alpha)z - i\beta) = \sum_{r=0}^{\infty} \sum_{m=1}^{\infty} \frac{e^{2\pi i m\beta}}{m} e^{-2\pi m(r+\alpha)z}$$

$$= \sum_{m=1}^{\infty} \frac{e^{2\pi i m\beta}}{m} e^{-2\pi m\alpha z} \sum_{r=0}^{\infty} e^{-2\pi m r z}$$

$$= \sum_{m=1}^{\infty} \frac{e^{2\pi i m\beta}}{m} \frac{e^{-2\pi m\alpha z}}{1 - e^{-2\pi m z}} = \sum_{m=1}^{\infty} e^{2\pi i m\beta} f_\alpha(m),$$

say, where

$$f_\alpha(m) = \frac{1}{m} \frac{e^{-2\pi m\alpha z}}{1 - e^{-2\pi m z}}.$$

As $m \to \infty$, $f_\alpha(m) \to 0$ *uniformly* in α if $0 \leq \alpha \leq 1$. Therefore the series

$$\sum_{m=1}^{\infty} e^{2\pi i m\beta} f_\alpha(m)$$

converges uniformly in α if $0 \leq \alpha \leq 1$, provided $0 < \beta < 1$, so we can pass to the limit $\alpha \to 0+$ term by term. This gives us

$$\lim_{\alpha \to 0+} \sum_{r=0}^{\infty} \lambda((r + \alpha)z - i\beta) = \sum_{r=0}^{\infty} \lambda(rz - i\beta).$$

Therefore, if $0 < \beta < 1$ we can let $\alpha \to 0+$ in the functional equation. The other limiting cases follow from the invariance of the formula under the following replacements:

$$\alpha \to 1 - \alpha, \qquad \beta \to 1 - \beta$$

$$\alpha \to \beta, \qquad \beta \to 1 - \alpha, \qquad z \to \frac{1}{z}$$

$$\alpha \to 1 - \beta, \qquad \beta \to \alpha, \qquad z \to \frac{1}{z}. \qquad \square$$

3.6 Deduction of Dedekind's functional equation from Iseki's formula

Now we use Iseki's formula to prove Equation (16) of Lemma 2. This, in turn, will prove Dedekind's functional equation for $\eta(\tau)$.

Equation (16) involves integers h and k with $k > 0$. First we treat the case $k = 1$ for which Equation (16) becomes

$$(26) \quad \sum_{n=1}^{\infty} \lambda\{n(z - ih)\} = \sum_{n=1}^{\infty} \lambda\left\{n\left(\frac{1}{z} - iH\right)\right\} + \frac{1}{2}\log z - \frac{\pi}{12}\left(z - \frac{1}{z}\right).$$

Since $\lambda(x)$ is periodic with period i this can be written as

$$(27) \quad \sum_{n=1}^{\infty} \lambda(nz) = \sum_{n=1}^{\infty} \lambda\left(\frac{n}{z}\right) + \frac{1}{2}\log z - \frac{\pi}{12}\left(z - \frac{1}{z}\right).$$

We can deduce this from Iseki's formula (18) by taking $\beta = 0$ and letting $\alpha \to 0+$. Before we let $\alpha \to 0+$ we separate the term $r = 0$ in the first term of the series on the left of (18) and in the second term of the series on the right of (18). The difference of these two terms is $\lambda(\alpha z) - \lambda(i\alpha)$. Each of these tends to ∞ as $\alpha \to 0+$ but their difference tends to a finite limit. We compute this limit as follows:

$$\lambda(\alpha z) - \lambda(i\alpha) = \log(1 - e^{-2\pi i\alpha}) - \log(1 - e^{-2\pi\alpha z}) = \log\frac{1 - e^{-2\pi i\alpha}}{1 - e^{-2\pi\alpha z}}.$$

By L'Hôpital's rule,

$$\lim_{\alpha \to 0+} \frac{1 - e^{-2\pi i\alpha}}{1 - e^{-2\pi\alpha z}} = \lim_{\alpha \to 0} \frac{2\pi i}{2\pi z} = \frac{i}{z}$$

so

$$\lim_{\alpha \to 0+} (\lambda(\alpha z) - \lambda(i\alpha)) = \log\frac{i}{z} = \frac{\pi i}{2} - \log z.$$

Now when $\alpha \to 0+$ the remaining terms in each series in (18) double up and we obtain, in the limit,

$$(28) \quad \frac{\pi i}{2} - \log z + 2\sum_{r=1}^{\infty} \lambda(rz) = 2\sum_{r=1}^{\infty} \lambda\left(\frac{r}{z}\right) - \frac{\pi z}{6} + \frac{\pi}{6z} + \frac{\pi i}{2}.$$

This reduces to (27) and proves (16) in the case $k = 1$.

Next we treat the case $k > 1$. We choose rational values for α and β in Iseki's formula (18) as follows. Take

$$\alpha = \frac{\mu}{k}, \quad \text{where } 1 \leq \mu \leq k - 1$$

and write

$$h\mu = qk + v, \quad \text{where } 1 \le v \le k - 1.$$

Now let

$$\beta = \frac{v}{k}.$$

Note that $v \equiv h\mu \pmod{k}$ so $-Hv \equiv -Hh\mu \equiv \mu \pmod{k}$, and therefore $-Hv/k \equiv \mu/k \pmod{1}$. Hence $\alpha = \mu/k \equiv -Hv/k \pmod{1}$ and $\beta = v/k \equiv h\mu/k \pmod{1}$. Substituting in Iseki's formula (18) and dividing by 2 we get

$$\frac{1}{2} \sum_{r=0}^{\infty} \left\{ \lambda\left(\left(r + \frac{\mu}{k}\right)z - i\frac{h\mu}{k}\right) + \lambda\left(\left(r + 1 - \frac{\mu}{k}\right)z + i\frac{h\mu}{k}\right) \right\}$$

$$= \frac{1}{2} \sum_{r=0}^{\infty} \left\{ \lambda\left(\left(r + \frac{v}{k}\right)\frac{1}{z} - i\frac{Hv}{k}\right) + \lambda\left(\left(r + 1 - \frac{v}{k}\right)\frac{1}{z} + i\frac{Hv}{k}\right) \right\}$$

$$- \frac{\pi z}{2}\left(\left(\frac{\mu}{k}\right)^2 - \frac{\mu}{k} + \frac{1}{6}\right) + \frac{\pi}{2z}\left(\left(\frac{v}{k}\right)^2 - \frac{v}{k} + \frac{1}{6}\right)$$

$$+ \pi i \left(\frac{\mu}{k} - \frac{1}{2}\right)\left(\frac{v}{k} - \frac{1}{2}\right).$$

Rewrite this as follows:

$$\frac{1}{2} \sum_{r=0}^{\infty} \left\{ \lambda\left(\frac{(rk + \mu)(z - ih)}{k}\right) + \lambda\left(\frac{(rk + k - \mu)(z - ih)}{k}\right) \right\}$$

$$= \frac{1}{2} \sum_{r=0}^{\infty} \left\{ \lambda\left(\frac{(rk + v)\left(\frac{1}{z} - iH\right)}{k}\right) + \lambda\left(\frac{(rk + k - v)\left(\frac{1}{z} - iH\right)}{k}\right) \right\}$$

$$- \frac{\pi z}{2}\left(\frac{\mu^2}{k^2} - \frac{\mu}{k} + \frac{1}{6}\right) + \frac{\pi}{2z}\left(\frac{v^2}{k^2} - \frac{v}{k} + \frac{1}{6}\right) + \pi i \left(\frac{\mu}{k} - \frac{1}{2}\right)\left(\frac{v}{k} - \frac{1}{2}\right).$$

Now sum both sides on μ for $\mu = 1, 2, \ldots, k - 1$ and note that

$$\{rk + \mu : r = 0, 1, 2, \ldots; \quad \mu = 1, 2, \ldots, k - 1\} = \{n : n \not\equiv 0 \pmod{k}\}$$

and similarly for the set of all numbers $rk + k - \mu$. Also, since $v \equiv h\mu \pmod{k}$, as μ runs through the numbers $1, 2, \ldots, k - 1$ then v runs through the same set of values in some other order. Hence we get

$$
\sum_{\substack{n=1 \\ n \not\equiv 0 \ (\mathrm{mod}\ k)}}^{\infty} \lambda\left(\frac{n}{k}(z-ih)\right) = \sum_{\substack{n=1 \\ n \not\equiv 0 \ (\mathrm{mod}\ k)}}^{\infty} \lambda\left(\frac{n}{k}\left(\frac{1}{z}-iH\right)\right) + \frac{\pi}{2}\left(\frac{1}{z}-z\right)\sum_{\mu=1}^{k-1}\frac{\mu^2}{k^2}
$$

$$
-\frac{\pi}{2}\left(\frac{1}{z}-z\right)\sum_{\mu=1}^{k-1}\frac{\mu}{k} + \frac{\pi}{12}\left(\frac{1}{z}-z\right)(k-1)
$$

$$
+\pi i\sum_{\mu=1}^{k-1}\frac{\mu}{k}\left(\frac{\nu}{k}-\frac{1}{2}\right) - \frac{\pi i}{2}\sum_{\mu=1}^{k-1}\frac{\nu}{k} + \frac{\pi i}{4}(k-1)
$$

$$
= \sum_{\substack{n=1 \\ n \not\equiv 0 \ (\mathrm{mod}\ k)}}^{\infty} \lambda\left(\frac{n}{k}\left(\frac{1}{z}-iH\right)\right) + \frac{\pi}{12}\left(\frac{1}{z}-z\right)
$$

$$
\times \left(\frac{(k-1)(2k-1)}{k} - 3(k-1) + (k-1)\right) + \pi i\sum_{\mu=1}^{k-1}\frac{\mu}{k}\left(\frac{\nu}{k}-\frac{1}{2}\right)
$$

$$
= \sum_{\substack{n=1 \\ n \not\equiv 0 \ (\mathrm{mod}\ k)}}^{\infty} \lambda\left(\frac{n}{k}\left(\frac{1}{z}-iH\right)\right) + \frac{\pi}{12}\left(z-\frac{1}{z}\right)\left(1-\frac{1}{k}\right) + \pi i\sum_{\mu=1}^{k-1}\frac{\mu}{k}\left(\frac{\nu}{k}-\frac{1}{2}\right).
$$

But ν was defined by the equation $h\mu = qk + \nu$, so we have

$$
\frac{h\mu}{k} = q + \frac{\nu}{k}, \qquad q = \left[\frac{h\mu}{k}\right], \qquad \frac{\nu}{k} = \frac{h\mu}{k} - \left[\frac{h\mu}{k}\right].
$$

Therefore

$$
\sum_{\mu=1}^{k-1}\frac{\mu}{k}\left(\frac{\nu}{k}-\frac{1}{2}\right) = \sum_{\mu=1}^{h-1}\frac{\mu}{k}\left(\frac{h\mu}{k}-\left[\frac{h\mu}{k}\right]-\frac{1}{2}\right) = s(h,k).
$$

Therefore we have proved that

$$
\text{(29)} \qquad \sum_{\substack{n=1 \\ n \not\equiv 0 \ (\mathrm{mod}\ k)}}^{\infty} \lambda\left(\frac{n}{k}(z-ih)\right) = \sum_{\substack{n=1 \\ n \not\equiv 0 \ (\mathrm{mod}\ k)}}^{\infty} \lambda\left(\frac{n}{k}\left(\frac{1}{z}-iH\right)\right)
$$

$$
+ \frac{\pi}{12}\left(z-\frac{1}{z}\right)\left(1-\frac{1}{k}\right) + \pi i s(h,k).
$$

Add this to Equation (27) which corresponds to the case $k = 1$:

$$
\sum_{m=1}^{\infty} \lambda(mz) = \sum_{m=1}^{\infty} \lambda\left(\frac{m}{z}\right) - \frac{\pi}{12}\left(z-\frac{1}{z}\right) + \frac{1}{2}\log z.
$$

This accounts for the missing terms in (29) with $n \equiv 0 \ (\mathrm{mod}\ k)$, if we write $n = mk$. When (27) is combined with (29) we get

$$
\sum_{n=1}^{\infty} \lambda\left(\frac{n}{k}(z-ih)\right) = \sum_{n=1}^{\infty} \lambda\left(\frac{n}{k}\left(\frac{1}{z}-iH\right)\right) - \frac{\pi}{12k}\left(z-\frac{1}{z}\right) + \frac{1}{2}\log z + \pi i s(h,k).
$$

This proves (16) which, in turn, completes the proof of Dedekind's functional equation for $\eta(\tau)$. For alternate proofs see p. 190 and [18], [35], and [45]. □

3.7 Properties of Dedekind sums

The Dedekind sums $s(h, k)$ which occur in the functional equation for $\eta(\tau)$ have applications to many parts of mathematics. Some of these are described in an excellent monograph on Dedekind sums by Rademacher and Grosswald [38]. We conclude this chapter with some arithmetical properties of the sums $s(h, k)$ which will be needed later in this book. In particular, Theorem 3.11 plays a central role in the study of the invariance of modular functions under transformations of certain subgroups of Γ, a topic discussed in the next chapter.

Note. Throughout this section we assume that k is a positive integer and that $(h, k) = 1$.

Dedekind sums are defined by the equation

$$(30) \qquad s(h, k) = \sum_{r=1}^{k-1} \frac{r}{k} \left(\frac{hr}{k} - \left[\frac{hr}{k} \right] - \frac{1}{2} \right).$$

First we express these sums in terms of the function $((x))$ defined by

$$((x)) = \begin{cases} x - [x] - \frac{1}{2} & \text{if } x \text{ is not an integer,} \\ 0 & \text{if } x \text{ is an integer.} \end{cases}$$

This is a periodic function of x with period 1, and $((-x)) = -((x))$. Actually, $((x))$ is the same as the Bernoulli periodic function $\bar{B}_1(x)$ discussed in [4], Chapter 12. Since $((x))$ is periodic and odd we find that

$$\sum_{r \bmod k} \left(\left(\frac{r}{k} \right) \right) = 0$$

and, more generally,

$$\sum_{r \bmod k} \left(\left(\frac{hr}{k} \right) \right) = 0 \quad \text{for } (h, k) = 1.$$

Since

$$\sum_{r \bmod k} \left(\left(\frac{r}{k} \right) \right) \left(\left(\frac{hr}{k} \right) \right) = \sum_{r=1}^{k} \left(\frac{r}{k} - \frac{1}{2} \right) \left(\left(\frac{hr}{k} \right) \right) = \sum_{r=1}^{k-1} \frac{r}{k} \left(\left(\frac{hr}{k} \right) \right)$$

the Dedekind sums can now be represented as follows:

$$(31) \qquad s(h, k) = \sum_{r \bmod k} \left(\left(\frac{r}{k} \right) \right) \left(\left(\frac{hr}{k} \right) \right).$$

This representation is often more convenient than (30) because we can exploit the periodicity of $((x))$.

61

Theorem 3.6

(a) *If* $h' \equiv \pm h$ (mod k), *then* $s(h', k) = \pm s(h, k)$, *with the same sign as in the congruence. Similarly, we have:*

(b) *If* $h\bar{h} \equiv \pm 1$ (mod k) *then* $s(\bar{h}, k) = \pm s(h, k)$.

(c) *If* $h^2 + 1 \equiv 0$ (mod k), *then* $s(h, k) = 0$.

PROOF. Parts (a) and (b) follow at once from (31). To prove (c) we note that $h^2 + 1 \equiv 0$ (mod k) implies $h \equiv -\bar{h}$ (mod k), where $\bar{h}$ is the reciprocal of h mod k, so from (a) and (b) we get $s(h, k) = -s(h, k) = 0$. $\qquad\square$

For small values of h the sum $s(h, k)$ can be easily evaluated from its definition. For example, when $h = 1$ we find

$$s(1, k) = \sum_{r=1}^{k-1} \frac{r}{k} \left(\frac{r}{k} - \frac{1}{2} \right) = \frac{1}{k^2} \sum_{r=1}^{k-1} r^2 - \frac{1}{2k} \sum_{r=1}^{k-1} r$$

$$= \frac{(k-1)(2k-1)}{6k} - \frac{k-1}{4} = \frac{(k-1)(k-2)}{12k}.$$

Similarly, the reader can verify that

$$s(2, k) = \frac{(k-1)(k-5)}{24k} \quad \text{if } k \text{ is odd.}$$

In general there is no simple formula for evaluating $s(h, k)$ in closed form. However, the sums satisfy a remarkable reciprocity law which can be used as an aid in calculating $s(h, k)$.

3.8 The reciprocity law for Dedekind sums

Theorem 3.7 (Reciprocity law for Dedekind sums). *If* $h > 0$, $k > 0$ *and* $(h, k) = 1$ *we have*

$$12hk s(h, k) + 12kh s(k, h) = h^2 + k^2 - 3hk + 1.$$

PROOF. Dedekind first deduced the reciprocity law from the functional equation for $\log \eta(\tau)$. We give an arithmetic proof of Rademacher and Whiteman [39], in which the sum $\sum_{r=1}^{k} ((hr/k))^2$ is evaluated in two ways. First we have

$$(32) \qquad \sum_{r=1}^{k} \left(\left(\frac{hr}{k} \right) \right)^2 = \sum_{r \bmod k} \left(\left(\frac{hr}{k} \right) \right)^2 = \sum_{r \bmod k} \left(\left(\frac{r}{k} \right) \right)^2 = \sum_{r=1}^{k-1} \left(\frac{r}{k} - \frac{1}{2} \right)^2.$$

We can also write

$$\sum_{r=1}^{k} \left(\left(\frac{hr}{k}\right)\right)^2 = \sum_{r=1}^{k-1} \left(\frac{hr}{k} - \left[\frac{hr}{k}\right] - \frac{1}{2}\right)^2$$

$$= \sum_{r=1}^{k-1} \left(\frac{h^2 r^2}{k^2} + \left[\frac{hr}{k}\right]^2 + \frac{1}{4} - \frac{hr}{k} + \left[\frac{hr}{k}\right] - \frac{2hr}{k}\left[\frac{hr}{k}\right]\right)$$

$$= 2h \sum_{r=1}^{k-1} \frac{r}{k}\left(\frac{hr}{k} - \left[\frac{hr}{k}\right] - \frac{1}{2}\right)$$

$$+ \sum_{r=1}^{k-1} \left[\frac{hr}{k}\right]\left(\left[\frac{hr}{k}\right] + 1\right) - \frac{h^2}{k^2}\sum_{r=1}^{k-1} r^2 + \frac{1}{4}\sum_{r=1}^{k-1} 1.$$

Comparing this with (32) and using (30) we obtain

$$(33) \qquad 2hs(h, k) + \sum_{r=1}^{k-1} \left[\frac{hr}{k}\right]\left(\left[\frac{hr}{k}\right] + 1\right) = \frac{h^2 + 1}{k^2}\sum_{r=1}^{k-1} r^2 - \frac{1}{k}\sum_{r=1}^{k-1} r.$$

In the sum on the left we collect those terms for which $[hr/k]$ has a fixed value. Since $0 < r < k$ we have $0 < hr/k < h$ and we can write

$$(34) \qquad \left[\frac{hr}{k}\right] = v - 1, \quad \text{where } v = 1, 2, \ldots, h.$$

For a given v let $N(v)$ denote the number of values of r for which $[hr/k] = v - 1$. Equation (34) holds if, and only if

$$v - 1 < \frac{hr}{k} < v, \qquad \text{or} \qquad \frac{k(v - 1)}{h} < r < \frac{kv}{h},$$

equality being excluded since $(h, k) = 1$ and $0 < r < k$. Therefore, if $1 \le v \le h - 1$, Equation (34) holds when r ranges from $[k(v - 1)/h] + 1$ to $[kv/h]$, and hence

$$N(v) = \left[\frac{kv}{h}\right] - \left[\frac{k(v - 1)}{h}\right] \quad \text{if } 1 \le v \le h - 1.$$

But when $v = h$ the quotient $kv/h = k$ and since $r = k$ is excluded we have

$$N(h) = k - 1 - \left[\frac{k(h - 1)}{h}\right].$$

Hence

$$(35) \quad \sum_{r=1}^{k-1} \left[\frac{hr}{k}\right]\left(\left[\frac{hr}{k}\right] + 1\right) = \sum_{v=1}^{h} (v - 1)vN(v)$$

$$= \sum_{v=1}^{h} (v - 1)v\left(\left[\frac{kv}{h}\right] - \left[\frac{k(v-1)}{h}\right]\right) - h(h-1)$$

$$= \sum_{v=1}^{h-1} \left[\frac{kv}{h}\right]\{(v-1)v - v(v+1)\}$$

$$+ kh(h-1) - h(h-1)$$

$$= -2\sum_{v=1}^{h-1} v\left[\frac{kv}{h}\right] + h(h-1)(k-1).$$

Now we also have

$$2hs(k, h) = 2\sum_{v=1}^{h-1} v\left(\frac{kv}{h} - \left[\frac{kv}{h}\right] - \frac{1}{2}\right) = -2\sum_{v=1}^{h-1} v\left[\frac{kv}{h}\right] + \frac{2k}{h}\sum_{v=1}^{h-1} v^2 - \sum_{v=1}^{h-1} v$$

so (35) becomes

$$\sum_{r=1}^{k-1} \left[\frac{hr}{k}\right]\left(\left[\frac{hr}{k}\right] + 1\right) = 2hs(k, h) - \frac{2k}{h}\sum_{v=1}^{h-1} v^2 + \sum_{v=1}^{h-1} v + h(h-1)(k-1).$$

We use this in (33) and multiply by $6k$ to obtain the reciprocity law. $\qquad\square$

3.9 Congruence properties of Dedekind sums

Theorem 3.8. *The number $6ks(h, k)$ is an integer. Moreover, if $\theta = (3, k)$ we have*

(a) $12hks(k, h) \equiv 0 \pmod{\theta k}$

and

(b) $12hks(h, k) \equiv h^2 + 1 \pmod{\theta k}$.

PROOF. From (30) we find

$$(36) \quad 6ks(h, k) = \frac{6h}{k}\sum_{r=1}^{k-1} r^2 - 6\sum_{r=1}^{k-1} r\left[\frac{hr}{k}\right] - 3\sum_{r=1}^{k-1} r.$$

Since $6\sum_{r=1}^{k-1} r^2 = k(k-1)(2k-1)$ each term on the right of (36) is an integer. Moreover, (36) shows that

$$6ks(h, k) \equiv h(k-1)(2k-1) \pmod 3$$

so we have

$$(37) \quad 12ks(h, k) \equiv 2h(k-1)(2k-1) \equiv h(k-1)(k+1) \pmod 3.$$

If $3|k$ then $3 \nmid h$ and (37) implies

$$12ks(h, k) \equiv -h \not\equiv 0 \pmod 3. \quad .$$

If $3 \nmid k$ then $3|(k - 1)(k + 1)$ and (37) implies

(38)
$$12ks(h, k) \equiv 0 \pmod 3.$$

In other words, $12ks(h, k) \equiv 0 \pmod 3$ if, and only if, $3 \nmid k$. Hence, interchanging h and k, we have

$$12hs(k, h) \equiv 0 \pmod 3 \quad \text{if, and only if, } 3 \nmid h.$$

If $\theta = 3$ this implies (a) since $(h, k) = 1$. If $\theta = 1$, (a) holds trivially. Part (a), together with the reciprocity law, gives (b) since $k^2 - 3hk \equiv 0 \pmod{\theta k}$.

$\square$

Note. Theorems 3.8(b) and 3.6(c) show that

$$s(h, k) = 0 \quad \text{if, and only if, } h^2 + 1 \equiv 0 \pmod k.$$

Theorem 3.9. *The Dedekind sums satisfy the congruence*

(39) $\quad 12ks(h, k) \equiv (k - 1)(k + 2) - 4h(k - 1) + 4 \displaystyle\sum_{r < k/2} \left[\frac{2hr}{k} \right] \pmod 8.$

If k is odd this becomes

(40)
$$12ks(h, k) \equiv k - 1 + 4 \sum_{r < k/2} \left[\frac{2hr}{k} \right] \pmod 8.$$

PROOF. From (36) we obtain

$$12ks(h, k) = 2h(k - 1)(2k - 1) - 12 \sum_{r=1}^{k-1} r \left[\frac{hr}{k} \right] - 3k(k - 1)$$

$$= -2h(k - 1) + 4hk(k - 1) - 12 \sum_{r=1}^{k-1} r \left[\frac{hr}{k} \right]$$

$$+ k(k - 1) - 4k(k - 1).$$

Now we reduce the right member modulo 8. Since $4k(k - 1) \equiv 0 \pmod 8$ this gives us

$$12ks(h, k) \equiv -2h(k - 1) - 4 \sum_{r=1}^{k-1} r \left[\frac{hr}{k} \right] + k(k - 1) \pmod 8$$

$$\equiv (k - 1)(k - 2h) - 4 \sum_{\substack{r=1 \\ r \text{ odd}}}^{k-1} \left[\frac{hr}{k} \right] \pmod 8$$

$$\equiv (k - 1)(k - 2h) - 4 \sum_{r=1}^{k-1} \left[\frac{hr}{k} \right] + 4 \sum_{r < k/2} \left[\frac{2hr}{k} \right] \pmod 8.$$

65

The next to last term is equal to

$$-4\sum_{r=1}^{k-1}\left[\frac{hr}{k}\right] = 4\sum_{r=1}^{k-1}\left(\left(\frac{hr}{k}\right)\right) - 4\sum_{r=1}^{k-1}\frac{hr}{k} + 2\sum_{r=1}^{k-1}1$$

$$= 0 - 2h(k-1) + 2(k-1) = (k-1)(2-2h).$$

Since

$$(k-1)(k-2h) + (k-1)(2-2h) = (k-1)(k+2) - 4h(k-1)$$

this proves (39).

When k is odd we have $4h(k-1) \equiv 0 \pmod{8}$ and

$$(k-1)(k+2) = k^2 + k - 2 \equiv k - 1 \pmod{8}$$

since $k^2 \equiv 1 \pmod{8}$. Hence (39) implies (40). □

Theorem 3.10. *If* $k = 2^\lambda k_1$ *where* $\lambda > 0$ *and* k_1 *is odd, then for odd* $h \geq 1$ *we have*

$$(41) \quad 12hks(h,k) \equiv h^2 + k^2 + 1 + 5k - 4k\sum_{v<h/2}\left[\frac{2kv}{h}\right] \pmod{2^{\lambda+3}}.$$

PROOF. Since h is odd we can apply (40) to obtain, after multiplication by k,

$$12hks(k,h) \equiv k(h-1) + 4k\sum_{v<h/2}\left[\frac{2kv}{h}\right] \pmod{2^{\lambda+3}}.$$

By the reciprocity law we have

$$12hks(h,k) = h^2 + k^2 - 3hk + 1 - 12hks(k,h)$$

$$\equiv h^2 + k^2 - 3hk + 1 - k(h-1) - 4k\sum_{v<h/2}\left[\frac{2kv}{h}\right] \pmod{2^{\lambda+3}}$$

$$\equiv h^2 + k^2 + 1 + k - 4hk - 4k\sum_{v<h/2}\left[\frac{2kv}{h}\right] \pmod{2^{\lambda+3}}.$$

Since h is odd we have $4k(h+1) \equiv 0 \pmod{2^{\lambda+3}}$ hence $k - 4hk \equiv 5k \pmod{2^{\lambda+3}}$ and we obtain (41). □

Finally, we obtain a property of Dedekind sums which plays a central role in the study of the invariance of modular functions under transformations of certain subgroups of the modular group. This will be needed in Chapter 4.

Theorem 3.11. *Let* $q = 3, 5, 7$ *or* 13 *and let* $r = 24/(q-1)$. *Given integers* a, b, c, d *with* $ad - bc = 1$ *such that* $c = c_1 q$, *where* $c_1 > 0$, *let*

$$\delta = \left\{s(a,c) - \frac{a+d}{12c}\right\} - \left\{s(a,c_1) - \frac{a+d}{12c_1}\right\}.$$

Then $r\delta$ *is an even integer.*

PROOF. Taking $k = c$ in Theorem 3.8(b) we find

$$12ac\left\{s(a, c) - \frac{a + d}{12c}\right\} \equiv a^2 + 1 - a(a + d) \equiv -bc \pmod{\theta c},$$

where $\theta = (3, c)$. The same theorem with $k = c_1 = c/q$ gives, after multiplication by q,

$$12ac\left\{s(a, c_1) - \frac{a + d}{12c_1}\right\} \equiv qa^2 + q - qa(a + d) \equiv -qbc \pmod{\theta_1 c},$$

where $\theta_1 = (3, c_1)$. Note that $\theta_1 | \theta$ so both congruences hold modulo $\theta_1 c$. Subtracting the congruences and multiplying by r we find

$$12acr\delta \equiv r(q - 1)bc \pmod{\theta_1 c}.$$

But $r(q - 1)bc = 24bc \equiv 0 \pmod{\theta_1 c}$ so this gives

$$12acr\delta \equiv 0 \pmod{\theta_1 c}.$$

Now $(a, c) = 1$ since $ad - bc = 1$. Also, $12c\delta$ is an integer so we can cancel a in the last congruence to get

(42) $$12cr\delta \equiv 0 \pmod{\theta_1 c}.$$

Next we show that we also have

(43) $$12cr\delta \equiv 0 \pmod{3c}.$$

Assume first that $q > 3$. In this case $\theta = (3, qc_1) = (3, c_1) = \theta_1$ so (42) becomes

$$12cr\delta \equiv 0 \pmod{\theta c}.$$

If $\theta = 3$ this gives (43). But if $\theta = 1$ then $3 \nmid c$ so $3 \nmid c_1$ and (38) implies $12cs(a, c) \equiv 0 \pmod 3$ and $12cs(a, c_1) \equiv 0 \pmod 3$. Hence

$$12cr\delta \equiv r(q - 1)(a + d) = 24(a + d) \equiv 0 \pmod 3,$$

which, together with (42), implies (43).

Now assume that $q = 3$ so $r = 12$. Then $\theta = 3$ and θ_1 is 1 or 3. If $\theta_1 = 3$ we get (43) by the same argument used above, so it remains to treat the case $\theta_1 = 1$. In this case $3 \nmid c_1$ so (38) implies $12c_1 s(a, c_1) \equiv 0 \pmod 3$, hence

$$12cs(a, c_1) \equiv 0 \pmod 9.$$

Also,

$$12c\delta = 12cs(a, c) - (a + d) - 12cs(a, c_1) + 3(a + d)$$
$$\equiv 12cs(a, c) + 2(a + d) \pmod 9,$$

so

(44) $$12rac\delta = 12racs(a, c) + 2r(a^2 + ad) \pmod 9.$$

But Theorem 3.8(b) gives us $12acs(a, c) \equiv a^2 + 1 \pmod 9$ since $3|c$. Hence (44) becomes

$$12rac\delta \equiv r(a^2 + 1) + 2ra^2 + 2rad \pmod 9$$
$$\equiv 3ra^2 + r + 2r(1 + bc) \equiv 3r + 2rbc \equiv 0 \pmod 9$$

since $r = 12$ and $9|12c$. This shows that

$$12rac\delta \equiv 0 \pmod 9.$$

Now $3\nmid a$ since $(a, c) = 1$ so we can cancel a to obtain $12rc\delta \equiv 0 \pmod 9$ which, with (42), implies (43).

Our next goal is to show that we also have

(45) $$12cr\delta \equiv 0 \pmod{24c}$$

since this implies $r\delta$ is even and proves the theorem. To prove (45) we treat separately the cases c odd and c even.

Case 1: *c odd.* Apply (40) with $k = c$ to obtain

$$12c\left\{s(a, c) - \frac{a + d}{12c}\right\} \equiv c - 1 + 4T(a, c) - (a + d) \pmod 8$$

where we have written

$$T(a, c) = \sum_{v < c/2} \left[\frac{2av}{c}\right].$$

We only need the fact that $T(a, c)$ is an integer. Applying (40) again with $k = c_1 = c/q$ and multiplying by q we have

$$12c\left\{s(a, c_1) - \frac{a + d}{12c_1}\right\} \equiv c - q + 4qT(a, c_1) - q(a + d) \pmod 8.$$

Subtracting the last two congruences and multiplying by r we find

$$12cr\delta \equiv r(q - 1) + r(q - 1)(a + d) \equiv 0 \pmod 8$$

since $r(q - 1) = 24$ and $4r \equiv 0 \pmod 8$. Combining this with (43) we obtain (45) and the theorem is proved for odd c.

Case 2: *c even.* Write $c = 2^\lambda \gamma$ with γ odd. Now a is odd since $(a, c) = 1$ so if $a \geq 1$ we can apply Theorem 3.10 with $k = c$ and $h = a$ to obtain

$$12ac\left\{s(a, c) - \frac{a + d}{12c}\right\} \equiv a^2 + c^2 + 1$$
$$+ 5c - 4cT(c, a) - a(a + d) \pmod{2^{\lambda + 3}}$$
$$\equiv c^2 + 5c - bc - 4cT(c, a) \pmod{2^{\lambda + 3}}$$

since $ad - bc = 1$. Similarly,

$$12ac\left\{s(a, c_1) - \frac{a + d}{12c_1}\right\} = cc_1 + 5c - qbc - 4cT(c_1, a) \pmod{2^{\lambda + 3}}.$$

Subtract, multiply by r and use the congruence $4cr \equiv 0 \pmod{2^{\lambda+3}}$ to obtain

$$12car\delta \equiv rcc_1(q-1) + r(q-1)bc \equiv 0 \pmod{2^{\lambda+3}}.$$

Since a is odd we can cancel a to obtain

(46) $$12cr\delta \equiv 0 \pmod{2^{\lambda+3}}.$$

Now (43) states that $12cr\delta \equiv 0 \pmod{3 \cdot 2^{\lambda}\gamma}$ which, together with (46) implies (45) and proves the theorem for $a \geq 1$.

To prove it for $a < 0$, write $\delta = \delta(a)$ to indicate the dependence on a. If $a' = a + tc$, where t is an integer, an easy calculation shows that $\delta(a') - \delta(a) = t(q-1)/12$ since $s(a, c) = s(a', c)$ and $s(a', c_1) = s(a, c_1)$. Therefore $r\delta(a') - r\delta(a) = 2t$, an even integer. Choosing t so that $a' \geq 1$ we know $r\delta(a')$ is even by the above argument, so $r\delta(a)$ is also even. This completes the proof. $\qquad\square$

3.10 The Eisenstein series $G_2(\tau)$

If k is an integer, $k \geq 2$, and if $\tau \in H$ the Eisenstein series

(47) $$G_{2k}(\tau) = \sum_{(m,n) \neq (0,0)} \frac{1}{(m+n\tau)^{2k}}$$

converges absolutely and has the Fourier expansion

(48) $$G_{2k}(\tau) = 2\zeta(2k) + \frac{2(2\pi i)^{2k}}{(2k-1)!} \sum_{n=1}^{\infty} \sigma_{2k-1}(n)e^{2\pi i n\tau}$$

where, as usual, $\sigma_{\alpha}(n) = \sum_{d|n} d^{\alpha}$. The cases $k = 2$ and $k = 3$ were worked out in detail in Chapter 1, and the same argument proves (48) for any $k \geq 2$. If $k = 1$ the series in (47) no longer converges absolutely. However, the series in (48) does converge absolutely and can be used to define the function $G_2(\tau)$.

Definition. If $\tau \in H$ we define

(49) $$G_2(\tau) = 2\zeta(2) + 2(2\pi i)^2 \sum_{n=1}^{\infty} \sigma(n)e^{2\pi i n\tau}.$$

If $x = e^{2\pi i \tau}$ the series on the right of (49) is an absolutely convergent power series for $|x| < 1$ so $G_2(\tau)$ is analytic in H. This definition also shows that $G_2(\tau + 1) = G_2(\tau)$.

Exercises 1 through 5 describe the behavior of G_2 under the other generator of the modular group. They show that

(50) $$G_2\left(\frac{-1}{\tau}\right) = \tau^2 G_2(\tau) - 2\pi i\tau,$$

a relation which leads to another proof of the functional equation $\eta(-1/\tau) = (-i\tau)^{1/2}\eta(\tau)$.

Exercises for Chapter 3

1. If $\tau \in H$ prove that

(51) $$G_2(\tau) = 2\zeta(2) + \sum_{\substack{n=-\infty \\ n \neq 0}}^{\infty} \sum_{m=-\infty}^{\infty} \frac{1}{(m + n\tau)^2}.$$

Hint: Start with Equation (12) of Chapter 1, replace τ by $n\tau$, where $n > 0$, and sum over all $n \geq 1$.

2. Use the series in (51) to show that

(52) $$\tau^{-2}G_2\left(\frac{-1}{\tau}\right) = 2\zeta(2) + \sum_{m=-\infty}^{\infty} \sum_{\substack{n=-\infty \\ n \neq 0}}^{\infty} \frac{1}{(m + n\tau)^2},$$

the iterated series in (52) being the same as that in (51) except with the order of summation reversed. Therefore, proving (50) is equivalent to showing that

(53) $$\sum_{m=-\infty}^{\infty} \sum_{\substack{n=-\infty \\ n \neq 0}}^{\infty} \frac{1}{(m + n\tau)^2} = \sum_{\substack{n=-\infty \\ n \neq 0}}^{\infty} \sum_{m=-\infty}^{\infty} \frac{1}{(m + n\tau)^2} - \frac{2\pi i}{\tau}.$$

3. (a) In the gamma function integral $\Gamma(z) = \int_0^\infty e^{-t}t^{z-1}\, dt$ make the change of variable $t = \alpha u$, where $\alpha > 0$, to obtain the formula

(54) $$\alpha^{-z}\Gamma(z) = \int_0^\infty e^{-\alpha u}u^{z-1}\, du,$$

and extend it by analytic continuation to complex α with $\mathrm{Re}(\alpha) > 0$.

(b) Take $z = 2$ and $\alpha = -2\pi i(m + n\tau)$ in (54) and sum over all $n \geq 1$ to obtain the relation

$$\sum_{\substack{n=-\infty \\ n \neq 0}}^{\infty} \frac{1}{(n\tau + m)^2} = -8\pi^2 \int_0^\infty \cos(2\pi mu)g_\tau(u)\, du,$$

where

$$g_\tau(u) = u \sum_{n=1}^{\infty} e^{2\pi i n\tau u} \quad \text{if } u > 0$$

and

$$g_\tau(0) = \lim_{u \to 0+} g_\tau(u) = \frac{-1}{2\pi i\tau}.$$

4. (a) Use Exercise 3 to deduce that

(55) $$\sum_{m=-\infty}^{\infty} \sum_{\substack{n=-\infty \\ n \neq 0}}^{\infty} \frac{1}{(n\tau + m)^2} = -8\pi^2 \sum_{m=-\infty}^{\infty} \int_0^1 f(t) \cos(2\pi mt)\, dt,$$

where

$$f(t) = \sum_{k=0}^{\infty} g_\tau(t + k).$$

(b) The series on the right of (55) is a Fourier series which converges to the value $\frac{1}{2}\{f(0+) + f(1-)\}$. Show that

$$f(0+) = \frac{-1}{2\pi i \tau} + \sum_{k=1}^{\infty} g_\tau(k)$$

and that

$$f(1-) = \sum_{k=1}^{\infty} g_\tau(k) = \sum_{n=1}^{\infty} \sigma(n)e^{2\pi i n\tau},$$

and then use (55) to obtain (50).

5. (a) Use the product defining $\eta(\tau)$ to show that

$$-4\pi i \frac{d}{d\tau} \log \eta(\tau) = G_2(\tau).$$

(b) Show that (50) implies

$$\frac{d}{d\tau} \log \eta\left(\frac{-1}{\tau}\right) = \frac{d}{d\tau} \log \eta(\tau) + \frac{1}{2}\frac{d}{d\tau} \log(-i\tau).$$

Integration of this equation gives $\eta(-1/\tau) = C(-i\tau)^{1/2}\eta(\tau)$ for some constant C. Taking $\tau = i$ we find $C = 1$.

6. Derive the reciprocity law for the Dedekind sums $s(h, k)$ from the transformation formula for $\log \eta(\tau)$ as given in Equation (12).

Exercises 7 and 8 describe properties of the function

$$\Phi(\alpha, \beta, s) = \frac{\Gamma(s)}{(2\pi)^s} \{\zeta(s, \alpha)F(\beta, 1 + s) + \zeta(s, 1 - \alpha)F(1 - \beta, 1 + s)\}$$

which occurs in the proof of Iseki's formula (Theorem 3.5). The properties follow from Hurwitz's formula (Theorem 12.6 of [4]) which states that

$$\zeta(1 - s, a) = \frac{\Gamma(s)}{(2\pi)^s} \{e^{-\pi i s/2}F(a, s) + e^{\pi i s/2}F(-a, s)\}.$$

7. (a) If $0 < a < 1$ and Re $(s) > 1$, prove that Hurwitz's formula implies

$$F(a, s) = \frac{\Gamma(1 - s)}{(2\pi)^{1-s}} \{e^{\pi i(1 - s)/2}\zeta(1 - s, a) + e^{\pi i(s - 1)/2}\zeta(1 - s, 1 - a)\}.$$

(b) Use (a) to show that $\Phi(\alpha, \beta, s)$ can be expressed in terms of Hurwitz zeta functions by the formula

$$\frac{\Phi(\alpha, \beta, s)}{\Gamma(s)\Gamma(-s)} = e^{\pi i s/2}\{\zeta(s, \alpha)\zeta(-s, 1 - \beta) + \zeta(s, 1 - \alpha)\zeta(-s, \beta)\}$$

$$+ e^{-\pi i s/2}\{\zeta(-s, 1 - \beta)\zeta(s, 1 - \alpha) + \zeta(-s, \beta)\zeta(s, \alpha)\}$$

and deduce that $\Phi(\alpha, \beta, s) = \Phi(1 - \beta, \alpha, -s)$.

8. This exercise gives an estimate for the modulus of the function $z^{-s}\Phi(\alpha, \beta, s)$ which occurs in the integral representation of $\Lambda(\alpha, \beta, s)$ in the proof of Iseki's formula (Theorem 3.5).

(a) Show that the formula of Exercise 7(b) implies

$$z^{-s}\Phi(\alpha, \beta, s) = \frac{-\pi z^{-s}}{s \sin \pi s} \{e^{-\pi i s/2}[\zeta(s, \alpha)\zeta(-s, \beta) + \zeta(s, 1 - \alpha)\zeta(-s, 1 - \beta)]$$
$$+ e^{\pi i s/2}[\zeta(s, \alpha)\zeta(-s, 1 - \beta) + \zeta(s, 1 - \alpha)\zeta(-s, \beta)]\}.$$

(b) For fixed z with $|\arg z| < \pi/2$, choose $\delta > 0$ so that $|\arg z| \leq \pi/2 - \delta$, and show that if $s = \sigma + it$ where $\sigma \geq -\frac{3}{2}$ we have

$$|z^{-s}| = O(e^{|t|(\pi/2 - \delta)}),$$

where the constant implied by the O-symbol depends on z.

(c) If $s = \sigma + it$ where $\sigma \geq -\frac{3}{2}$ and $|t| \geq 1$, show that

$$\frac{1}{|s \sin \pi s|} = O\left(\frac{e^{-\pi |t|}}{|t|}\right),$$

and that

$$|e^{\pi i s/2}| = O(e^{\pi |t|/2}), \qquad |e^{-\pi i s/2}| = O(e^{\pi |t|/2}).$$

(d) If $\sigma \geq -\frac{3}{2}$ and $|t| \geq 1$ obtain the estimate $|\zeta(s, a)| = O(|t|^c)$ for some $c > 0$ (see [4], Theorem 12.23) and use (b) to deduce that

$$|z^{-s}\Phi(\alpha, \beta, s)| = O(|t|^{2c-1}e^{-|t|\delta}).$$

This shows that the integral of $z^{-s}\Phi(\alpha, \beta, s)$ along the horizontal segments of the rectangle in Figure 3.2 tends to 0 as $T \to \infty$.

PROPERTIES OF DEDEKIND SUMS

9. If $k \geq 1$ the equation

$$s(h, k) = \sum_{r \bmod k} \left(\left(\frac{r}{k}\right)\right)\left(\left(\frac{hr}{k}\right)\right)$$

is meaningful even if h is not relatively prime to k and is sometimes taken as the definition of Dedekind sums. Using this as the definition of $s(h, k)$ prove that $s(qh, qk) = s(h, k)$ if $q > 0$.

10. If p is prime prove that

$$(p + 1)s(h, k) = s(ph, k) + \sum_{m=0}^{p-1} s(h + mk, pk).$$

11. For integers r, h, k with $k \geq 1$ prove that we have the finite Fourier expansion

$$\left(\left(\frac{hr}{k}\right)\right) = -\frac{1}{2k} \sum_{v=1}^{k-1} \sin \frac{2\pi hrv}{k} \cot \frac{\pi v}{k}$$

and derive the following expression for Dedekind sums:

$$s(h, k) = \frac{1}{4k} \sum_{r=1}^{k-1} \cot \frac{\pi hr}{k} \cot \frac{\pi r}{k}.$$

12. This exercise relates Dedekind sums with the sequence $\{u(n)\}$ of Fibonacci numbers $1, 1, 2, 3, 5, 8, \ldots$, in which $u(1) = u(2) = 1$ and $u(n + 1) = u(n) + u(n - 1)$.
(a) If $h = u(2n)$ and $k = u(2n + 1)$ prove that $s(h, k) = 0$.
(b) If $h = u(2n - 1)$ and $k = u(2n)$ prove that $12hks(h, k) = h^2 + k^2 - 3hk + 1$.

The following exercises give a number of formulas for evaluating Dedekind sums in closed form in special cases. Assume throughout that $(h, k) = 1$, $k \geq 1, h \geq 1$.

13. If $k \equiv r \pmod{h}$ prove that the reciprocity law implies

$$12hk\,s(h, k) = k^2 - \{12s(r, h) + 3\}hk + h^2 + 1.$$

Use the result of Exercise 13 to deduce the following formulas:

14. If $k \equiv 1 \pmod{h}$ then $12hks(h, k) = (k - 1)(k - h^2 - 1)$.

15. If $k \equiv 2 \pmod{h}$ then $12hks(h, k) = (k - 2)(k - \frac{1}{2}(h^2 + 1))$.

16. If $k \equiv -1 \pmod{h}$ then $12hks(h, k) = k^2 + (h^2 - 6h + 2)k + h^2 + 1$.

17. If $k \equiv r \pmod{h}$ and if $h \equiv t \pmod{r}$ where $r \geq 1$ and $t = \pm 1$, then

$$12hk\,s(h, k) = k^2 - \frac{h^2 - t(r - 1)(r - 2)h + r^2 + 1}{r}k + h^2 + 1.$$

This formula includes those of Exercises 14 and 15 as special cases.

18. Show that the formula of Exercise 17 determines $s(h, k)$ completely when $r = 3$ and when $r = 4$.

19. If $k \equiv 5 \pmod{h}$ and if $h \equiv t \pmod 5$, where $t = \pm 1$ or ± 2, then

$$12hk\,s(h, k) = k^2 - \frac{h^2 + 4t(t - 2)(t + 2)h + 26}{5}k + h^2 + 1.$$

20. Assume $0 < h < k$ and let $r_0, r_1, \ldots, r_{n+1}$ denote the sequence of remainders in the Euclidean algorithm for calculating the gcd (h, k), so that

$$r_0 = k, \quad r_1 = h, \quad r_{j+1} \equiv r_{j-1} \pmod{r_j}, \quad 1 \leq r_{j+1} < r_j, \quad r_{n+1} = 1.$$

Prove that

$$s(h, k) = \frac{1}{12} \sum_{j=1}^{n+1} \left\{ (-1)^{j+1} \frac{r_j^2 + r_{j-1}^2 + 1}{r_j r_{j-1}} \right\} - \frac{(-1)^n + 1}{8}.$$

This also expresses $s(h, k)$ as a finite sum, but with fewer terms than the sum in the original definition.

4

Congruences for the coefficients of the modular function j

4.1 Introduction

The function $j(\tau) = 12^3 J(\tau)$ has a Fourier expansion of the form

$$j(\tau) = \frac{1}{x} + \sum_{n=0}^{\infty} c(n)x^n, \quad (x = e^{2\pi i \tau})$$

where the coefficients $c(n)$ are integers. At the end of Chapter 1 we mentioned a number of congruences involving these integers. This chapter shows how some of these congruences are obtained. Specifically we will prove that

$$c(2n) \equiv 0 \pmod{2^{11}},$$
$$c(3n) \equiv 0 \pmod{3^5},$$
$$c(5n) \equiv 0 \pmod{5^2},$$
$$c(7n) \equiv 0 \pmod{7}.$$

The method used to obtain these congruences can be illustrated for the modulus 5^2. We consider the function

$$f_5(\tau) = \sum_{n=1}^{\infty} c(5n)x^n$$

obtained by extracting every fifth coefficient in the Fourier expansion of j. Then we show that there is an identity of the form

(1) $$f_5(\tau) = 25\{a_1 \Phi(\tau) + a_2 \Phi^2(\tau) + \cdots + a_k \Phi^k(\tau)\},$$

where the a_i are integers and $\Phi(\tau)$ has a power series expansion in $x = e^{2\pi i \tau}$ with integer coefficients. By equating coefficients in (1) we see that each coefficient of $f_5(\tau)$ is divisible by 25.

Success in this method depends on showing that such identities exist. How are they obtained?

74

Theorem 2.8 tells us that every modular function f is a rational function of j. Sometimes this rational function is a polynomial in j with integer coefficients, giving us an identity of the form

$$f(\tau) = a_1 j(\tau) + a_2 j^2(\tau) + \cdots + a_k j^k(\tau).$$

However, the function $f_5(\tau)$ is not invariant under all transformations of the modular group Γ and cannot be so expressed in terms of $j(\tau)$. But we shall find that $f_5(\tau)$ is invariant under the transformations of a certain subgroup of Γ, and the general theory enables us to express $f_5(\tau)$ as a polynomial in another basic function $\Phi(\tau)$ which plays the same role as $j(\tau)$ relative to this subgroup. This representation leads to an identity such as (1) and hence to the desired congruence property.

The subgroup in question is the set of all unimodular matrices $\begin{pmatrix} a & b \\ c & d \end{pmatrix}$ with $c \equiv 0 \pmod 5$. More generally we shall consider those matrices in Γ with $c \equiv 0 \pmod q$, where q is a prime or a power of a prime.

4.2 The subgroup $\Gamma_0(q)$

Definition. If q is any positive integer we define $\Gamma_0(q)$ to be the set of all matrices $\begin{pmatrix} a & b \\ c & d \end{pmatrix}$ in Γ with $c \equiv 0 \pmod q$.

It is easy to verify that $\Gamma_0(q)$ is a subgroup of Γ. The next theorem gives a way of representing each element of Γ in terms of elements of $\Gamma_0(p)$ when p is prime. In the language of group theory it shows that $\Gamma_0(p)$ is of finite index in Γ.

Theorem 4.1. *Let $S\tau = -1/\tau$ and $T\tau = \tau + 1$ be the generators of the full modular group Γ, and let p be any prime. Then for every V in Γ, $V \notin \Gamma_0(p)$, there exists an element P in $\Gamma_0(p)$ and an integer k, $0 \leq k < p$, such that*

$$V = PST^k.$$

PROOF. Given $V = \begin{pmatrix} A & B \\ C & D \end{pmatrix}$ where $C \not\equiv 0 \pmod p$. We wish to find

$$P = \begin{pmatrix} a & b \\ c & d \end{pmatrix}, \text{ with } c \equiv 0 \pmod p,$$

and an integer k, $0 \leq k < p$, such that

$$\begin{pmatrix} A & B \\ C & D \end{pmatrix} = \begin{pmatrix} a & b \\ c & d \end{pmatrix} ST^k = \begin{pmatrix} a & b \\ c & d \end{pmatrix} \begin{pmatrix} 0 & -1 \\ 1 & 0 \end{pmatrix} \begin{pmatrix} 1 & k \\ 0 & 1 \end{pmatrix} = \begin{pmatrix} a & b \\ c & d \end{pmatrix} \begin{pmatrix} 0 & -1 \\ 1 & k \end{pmatrix}.$$

75

All matrices here are nonsingular so we can solve for $\begin{pmatrix} a & b \\ c & d \end{pmatrix}$ to get

$$\begin{pmatrix} a & b \\ c & d \end{pmatrix} = \begin{pmatrix} A & B \\ C & D \end{pmatrix}\begin{pmatrix} 0 & -1 \\ 1 & k \end{pmatrix}^{-1} = \begin{pmatrix} A & B \\ C & D \end{pmatrix}\begin{pmatrix} k & 1 \\ -1 & 0 \end{pmatrix} = \begin{pmatrix} kA - B & A \\ kC - D & C \end{pmatrix}.$$

Choose k to be that solution of the congruence

$$kC \equiv D \pmod{p} \quad \text{with } 0 \leq k < p.$$

This is possible since $C \not\equiv 0 \pmod{p}$. Now take

$$c = kC - D, \qquad a = kA - B, \qquad b = A, \qquad d = C.$$

Then $c \equiv 0 \pmod{p}$ so $P \in \Gamma_0(p)$. This completes the proof. $\qquad\square$

4.3 Fundamental region of $\Gamma_0(p)$

As usual we write $S\tau = -1/\tau$ and $T\tau = \tau + 1$, and let R_Γ denote the fundamental region of Γ.

Theorem 4.2. *For any prime p the set*

$$R_\Gamma \cup \bigcup_{k=0}^{p-1} ST^k(R_\Gamma)$$

is a fundamental region of the subgroup $\Gamma_0(p)$.

This theorem is illustrated for $p = 3$ in Figure 4.1.

PROOF. Let R denote the set

$$R = R_\Gamma \cup \bigcup_{k=0}^{p-1} ST^k(R_\Gamma).$$

We will prove

(i) if $\tau \in H$, there is a V in $\Gamma_0(p)$ such that $V\tau$ belongs to the closure of R, and
(ii) no two distinct points of R are equivalent under $\Gamma_0(p)$.

To prove (i), choose τ in H, choose τ_1 in the closure of R_Γ and choose A in Γ such that $A\tau = \tau_1$. Then by Theorem 4.1 we can write

$$A^{-1} = PW$$

where $P \in \Gamma_0(p)$ and $W = I$ or $W = ST^k$ for some k, $0 \leq k \leq p - 1$. Then $P = A^{-1}W^{-1}$ and $P^{-1} = WA$. Let $V = P^{-1}$. Then $V \in \Gamma_0(p)$ and

$$V\tau = WA\tau = W\tau_1.$$

Since $W = I$ or $W = ST^k$, this proves (i).

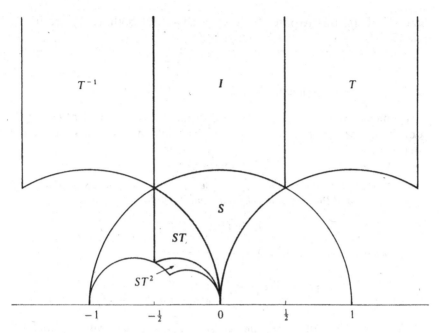

Figure 4.1   Fundamental region for $\Gamma_0(3)$

Next we prove (ii). Suppose $\tau_1 \in R$, $\tau_2 \in R$ and $V\tau_1 = \tau_2$ for some V in $\Gamma_0(p)$. We will prove that $\tau_1 = \tau_2$. There are three cases to consider:

(a) $\tau_1 \in R_\Gamma$, $\tau_2 \in R_\Gamma$. In this case $\tau_1 = \tau_2$ since $V \in \Gamma$.
(b) $\tau_1 \in R_\Gamma$, $\tau_2 \in ST^k(R_\Gamma)$.
(c) $\tau_1 \in ST^{k_1}(R_\Gamma)$, $\tau_2 \in ST^{k_2}(R_\Gamma)$.

In case (b), $\tau_2 = ST^k\tau_3$ where $\tau_3 \in R_\Gamma$. The equation

$$V\tau_1 = \tau_2 \quad \text{implies} \quad V\tau_1 = ST^k\tau_3, \tau_1 = V^{-1}ST^k\tau_3.$$

But $\tau_1 \in R_\Gamma$ and $\tau_3 \in R_\Gamma$ so $V^{-1}ST^k = I$,

$$V = ST^k = \begin{pmatrix} 0 & -1 \\ 1 & k \end{pmatrix}.$$

This contradicts the fact that $V \in \Gamma_0(p)$.

Finally, consider case (c). In this case

$$\tau_1 = ST^{k_1}\tau_1' \quad \text{and} \quad \tau_2 = ST^{k_2}\tau_2'$$

where τ_1' and τ_2' are in R_Γ. Since $V\tau_1 = \tau_2$ we have $VST^{k_1}\tau_1' = ST^{k_2}\tau_2'$ so $VST^{k_1} = ST^{k_2}$,

$$V = ST^{k_2-k_1}S = \begin{pmatrix} -1 & 0 \\ k_2 - k_1 & -1 \end{pmatrix}.$$

Since $V \in \Gamma_0(p)$ this requires $k_2 \equiv k_1 \pmod{p}$. But both k_1, k_2 are in the interval $[0, p - 1]$, so $k_2 = k_1$. Therefore

$$V = ST^0S = S^2 = I$$

and $\tau_1 = \tau_2$. This completes the proof. □

We mention (without proof) the following theorem of Rademacher [34] concerning the generators of $\Gamma_0(p)$. (This theorem is not needed in the later work.)

Theorem 4.3. *For any prime $p > 3$ the subgroup $\Gamma_0(p)$ has $2[p/12] + 3$ generators and they may be selected from the following elements:*

$$T, V_1, V_2, \ldots, V_{p-1},$$

where $T\tau = \tau + 1$, $S\tau = -1/\tau$, and

$$V_k = ST^kST^{-k'}S = \begin{pmatrix} k' & 1 \\ -(kk' + 1) & -k \end{pmatrix},$$

where $kk' \equiv -1 \pmod{p}$. The subgroup $\Gamma_0(2)$ has generators T and V_1: the subgroup $\Gamma_0(3)$ has generators T and V_2.

Here is a short table of generators:

p	2	3	5	7	11	13	17	19
Generators:	T	T	T	T	T	T	T	T
	V_1	V_2	V_2	V_3	V_4	V_4	V_4	V_5
			V_3	V_5	V_6	V_5	V_7	V_8
						V_8	V_9	V_{12}
						V_{10}	V_{13}	V_{13}

4.4 Functions automorphic under the subgroup $\Gamma_0(p)$

We recall that a modular function f is one which has the following three properties:

(a) f is meromorphic in the upper half-plane H.
(b) $f(A\tau) = f(\tau)$ for every transformation A in the modular group Γ.
(c) The Fourier expansion of f has the form

$$f(\tau) = \sum_{n=-m}^{\infty} a_n e^{2\pi i n \tau}.$$

If property (b) is replaced by

(b') $f(V\tau) = f(\tau)$ for every transformation V in $\Gamma_0(p)$,

then f is said to be *automorphic* under the subgroup $\Gamma_0(p)$. We also say that f belongs to $\Gamma_0(p)$.

The next theorem shows that the only bounded functions belonging to $\Gamma_0(p)$ are constants.

Theorem 4.4. *If f is automorphic under $\Gamma_0(p)$ and bounded in H, then f is constant.*

PROOF. According to Theorem 4.1, for every V in Γ there exists an element P in $\Gamma_0(p)$ and an integer k, $0 \leq k \leq p$, such that

$$V = PA_k,$$

where $A_k = ST^k$ if $k < p$, and $A_p = I$. For each $k = 0, 1, \ldots, p$, let

$$\Gamma_k = \{PA_k : P \in \Gamma_0(p)\}.$$

Each set Γ_k is called a *right coset* of $\Gamma_0(p)$. Choose an element V_k from the coset Γ_k and define a function f_k on H by the equation

$$f_k(\tau) = f(V_k \tau).$$

Note that $f_p(\tau) = f(P\tau) = f(\tau)$ since $P \in \Gamma_0(p)$ and f is automorphic under $\Gamma_0(p)$. The function value $f_k(\tau)$ does not depend on which element V_k was chosen from the coset Γ_k because

$$f_k(\tau) = f(V_k \tau) = f(PA_k \tau) = f(A_k \tau)$$

and the element A_k is the same for all members of the coset Γ_k.

How does f_k behave under the transformations of the full modular group? If $V \in \Gamma$ then

$$f_k(V\tau) = f(V_k V\tau).$$

Now $V_k V \in \Gamma$ so there is an element Q in $\Gamma_0(p)$ and an integer m, $0 \leq m \leq p$, such that

$$V_k V = QA_m.$$

Therefore we have

$$f_k(V\tau) = f(V_k V\tau) = f(QA_m \tau) = f(A_m \tau) = f_m(\tau).$$

Moreover, as k runs through the integers $0, 1, 2, \ldots, p$ so does m. In other words, there is a permutation σ of $\{0, 1, 2, \ldots, p\}$ such that

$$f_k(V\tau) = f_{\sigma(k)}(\tau) \quad \text{for each } k = 0, 1, \ldots, p.$$

Now choose a fixed w in H and let

$$\varphi(\tau) = \prod_{k=0}^{p} \{f_k(\tau) - f(w)\}.$$

79

Then if $V \in \Gamma$ we have

$$\varphi(V\tau) = \prod_{k=0}^{p} \{f_k(V\tau) - f(w)\} = \prod_{k=0}^{p} \{f_{\sigma(k)}(\tau) - f(w)\} = \varphi(\tau),$$

so φ is automorphic under the full group Γ. Now φ is bounded in H (since each f_k is). Therefore, φ omits some value hence, by Theorem 2.5, φ is constant, so $\varphi(\tau) = \varphi(w)$ for all τ. But $\varphi(w) = 0$ because

$$\varphi(w) = \prod_{k=0}^{p} \{f_k(w) - f(w)\}$$

and the factor with $k = p$ vanishes since $f_p = f$. Therefore $\varphi(\tau) = 0$ for all τ. Now take $\tau = i$. Then

$$0 = \prod_{k=0}^{p} \{f_k(i) - f(w)\}$$

hence some factor is 0. In other words, $f(w) = f_k(i)$ for some k. But w was arbitrary so f can take only the values $f_0(i), \ldots, f_p(i)$. This implies that f is constant. □

4.5 Construction of functions belonging to $\Gamma_0(p)$

This section shows how to construct functions automorphic under the subgroup $\Gamma_0(p)$ from given functions automorphic under Γ.

Theorem 4.5. *If f is automorphic under Γ and if p is prime, let*

$$f_p(\tau) = \frac{1}{p} \sum_{\lambda=0}^{p-1} f\left(\frac{\tau + \lambda}{p}\right).$$

Then f_p is automorphic under $\Gamma_0(p)$. Moreover, if f has the Fourier expansion

$$f(\tau) = \sum_{n=-m}^{\infty} a(n)e^{2\pi i n\tau}$$

then f_p has the Fourier expansion

$$f_p(\tau) = \sum_{n=-[m/p]}^{\infty} a(np)e^{2\pi i n\tau}.$$

PROOF. First we prove the statement concerning Fourier expansions. We have

$$f_p(\tau) = \frac{1}{p} \sum_{\lambda=0}^{p-1} \sum_{n=-m}^{\infty} a(n)e^{2\pi i n(\tau + \lambda)/p}$$

$$= \frac{1}{p} \sum_{n=-m}^{\infty} a(n)e^{2\pi i n\tau/p} \sum_{\lambda=0}^{p-1} e^{2\pi i n\lambda/p}.$$

But

$$\sum_{\lambda=0}^{p-1} e^{2\pi i n\lambda/p} = \begin{cases} 0 & \text{if } p \nmid n \\ p & \text{if } p \mid n \end{cases}$$

so

$$f_p(\tau) = \sum_{\substack{n=-m \\ p \mid n}}^{\infty} a(n)e^{2\pi i n\tau/p} = \sum_{n=-[m/p]}^{\infty} a(np)e^{2\pi i n\tau}.$$

This shows that f_p has the proper behavior at the point $\tau = i\infty$. Also, f_p is clearly meromorphic in H because it is a linear combination of functions meromorphic in H.

Next we must show that

$$f_p(V\tau) = f_p(\tau) \quad \text{whenever } V \in \Gamma_0(p).$$

For this we use a lemma.

Lemma 1. *If $V \in \Gamma_0(p)$ and if $0 \le \lambda \le p - 1$, let $T_\lambda\tau = (\tau + \lambda)/p$. Then there exists an integer μ, $0 \le \mu \le p - 1$ and a transformation W_μ in $\Gamma_0(p^2)$ such that*

$$T_\lambda V = W_\mu T_\mu.$$

Moreover, as λ runs through a complete residue system modulo p, so does μ.

First we use the lemma to complete the proof of Theorem 4.5, then we return to the proof of the lemma.

If $V \in \Gamma_0(p)$ we have

$$f_p(V\tau) = \frac{1}{p}\sum_{\lambda=0}^{p-1} f\left(\frac{V\tau + \lambda}{p}\right) = \frac{1}{p}\sum_{\lambda=0}^{p-1} f(T_\lambda V\tau).$$

Now we use the lemma to write the last sum as

$$\frac{1}{p}\sum_{\mu=0}^{p-1} f(W_\mu T_\mu\tau) = \frac{1}{p}\sum_{\mu=0}^{p-1} f(T_\mu\tau) = f_p(\tau).$$

This proves that f_p is invariant under all transformations in $\Gamma_0(p)$, so f_p is automorphic under $\Gamma_0(p)$. $\qquad\square$

PROOF OF LEMMA 1. Let $V = \begin{pmatrix} a & b \\ c & d \end{pmatrix}$, where $c \equiv 0 \pmod{p}$, and let λ be given, $0 \le \lambda \le p - 1$. We are to find an integer μ, $0 \le \mu \le p - 1$ and a transformation $W_\mu = \begin{pmatrix} A & B \\ C & D \end{pmatrix}$ such that $W_\mu \in \Gamma_0(p^2)$ and

$$T_\lambda V = W_\mu T_\mu.$$

Since $T_\lambda = \begin{pmatrix} 1 & \lambda \\ 0 & p \end{pmatrix}$ we must satisfy the matrix equation

$$\begin{pmatrix} 1 & \lambda \\ 0 & p \end{pmatrix}\begin{pmatrix} a & b \\ c & d \end{pmatrix} = \begin{pmatrix} A & B \\ C & D \end{pmatrix}\begin{pmatrix} 1 & \mu \\ 0 & p \end{pmatrix},$$

or

$$\begin{pmatrix} a + \lambda c & b + \lambda d \\ pc & pd \end{pmatrix} = \begin{pmatrix} A & A\mu + Bp \\ C & C\mu + Dp \end{pmatrix}$$

with $C \equiv 0 \pmod{p^2}$. Equating entries we must satisfy the relations

(2)
$$\begin{cases} A = a + \lambda c \\ C = pc \end{cases}$$

(3)
$$\begin{cases} A\mu + Bp = b + \lambda d \\ C\mu + Dp = pd \end{cases}$$

with

$$C \equiv 0 \pmod{p^2} \quad \text{and} \quad AD - BC = 1.$$

Now (2) determines A and C. Since $p|c$, we have $C \equiv 0 \pmod{p^2}$. Substituting these values in (3) we must satisfy

(4)
$$\begin{cases} (a + \lambda c)\mu + Bp = b + \lambda d \\ cp\mu + Dp = pd. \end{cases}$$

Choose μ to be that solution of the congruence

$$\mu a \equiv b + \lambda d \pmod{p}$$

which lies in the interval $0 \le \mu \le p - 1$. This is possible because $ad - bc = 1$ and $p|c$ imply $p \nmid a$. Note that distinct values of $\lambda \bmod p$ give rise to distinct values of $\mu \bmod p$. Then, since $p|c$ we have

$$\mu a + \mu \lambda c \equiv b + \lambda d \pmod{p}$$

or

$$(a + \lambda c)\mu \equiv b + \lambda d \pmod{p}.$$

Therefore there is an integer B such that

$$(a + \lambda c)\mu + Bp = b + \lambda d.$$

Therefore the first relation in (4) is satisfied. The second relation requires $D = d - c\mu$. Thus, we have found integers μ, A, B, C, D such that

$$\begin{pmatrix} 1 & \lambda \\ 0 & p \end{pmatrix}\begin{pmatrix} a & b \\ c & d \end{pmatrix} = \begin{pmatrix} A & B \\ C & D \end{pmatrix}\begin{pmatrix} 1 & \mu \\ 0 & p \end{pmatrix}.$$

Clearly $AD - BC = 1$ since all matrices in this equation have determinant 1 or p. This completes the proof of the lemma. $\square$

4.6 The behavior of f_p under the generators of Γ

Let $T\tau = \tau + 1$ and $S\tau = -1/\tau$ be the generators of Γ. Since $T \in \Gamma_0(p)$ we have $f_p(T\tau) = f_p(\tau)$. The next theorem gives a companion result for $f_p(S\tau)$.

Theorem 4.6. *If f is automorphic under Γ and if p is prime, then*

$$f_p\left(-\frac{1}{\tau}\right) = f_p(\tau) + \frac{1}{p} f(p\tau) - \frac{1}{p} f\left(\frac{\tau}{p}\right).$$

To prove this we need another lemma.

Lemma 2. *Let $T_\lambda \tau = (\tau + \lambda)/p$. Then for each λ in the interval $1 \leq \lambda \leq p - 1$ there exists an integer μ in the same interval and a transformation V in $\Gamma_0(p)$ such that*

$$T_\lambda S = V T_\mu.$$

Moreover, as λ runs through the numbers $1, 2, \ldots, p - 1$, so does μ.

PROOF OF LEMMA 2. We wish to find $\begin{pmatrix} a & b \\ c & d \end{pmatrix}$ in $\Gamma_0(p)$ such that

$$\begin{pmatrix} 1 & \lambda \\ 0 & p \end{pmatrix}\begin{pmatrix} 0 & -1 \\ 1 & 0 \end{pmatrix} = \begin{pmatrix} a & b \\ c & d \end{pmatrix}\begin{pmatrix} 1 & \mu \\ 0 & p \end{pmatrix}$$

or

$$\begin{pmatrix} \lambda & -1 \\ p & 0 \end{pmatrix} = \begin{pmatrix} a & a\mu + bp \\ c & c\mu + dp \end{pmatrix}.$$

Take $a = \lambda$, $c = p$ and let μ be that solution of the congruence

$$\lambda\mu \equiv -1 \pmod{p}$$

in the interval $1 \leq \mu \leq p - 1$. This solution is unique and μ runs through a reduced residue system mod p with λ. Choose b to be that integer such that $a\mu + bp = -1$, and take $d = -\mu$. Then $c\mu + dp = 0$ and the proof is complete. $\qquad\square$

PROOF OF THEOREM 4.6. We have

$$pf_p\left(-\frac{1}{\tau}\right) = \sum_{\lambda=0}^{p-1} f\left(\frac{S\tau + \lambda}{p}\right) = f\left(\frac{S\tau}{p}\right) + \sum_{\lambda=1}^{p-1} f(T_\lambda S\tau)$$

$$= f\left(-\frac{1}{\tau p}\right) + \sum_{\mu=1}^{p-1} f(V T_\mu \tau) = f(\tau p) + \sum_{\mu=0}^{p-1} f(T_\mu \tau) - f\left(\frac{\tau}{p}\right)$$

$$= f(\tau p) + pf_p(\tau) - f\left(\frac{\tau}{p}\right). \qquad\square$$

83

4.7 The function $\varphi(\tau) = \Delta(q\tau)/\Delta(\tau)$

The number of poles of an automorphic function in the closure of its fundamental region is called its *valence*. A function is called *univalent* on a subgroup G if it is automorphic under G and has valence 1. Such a function plays the same role in G that J plays in the full group Γ.

It can be shown (using Riemann surfaces) that univalent functions exist on G if and only if the genus of the fundamental region R_G is zero. [This is the topological genus of the surface obtained by identifying congruent edges of R_G. For example, the genus of R_Γ is zero because R_Γ is topologically equivalent to a sphere when its congruent edges are identified.]

Our next goal is to construct a univalent function on the subgroup $\Gamma_0(p)$ whenever the genus of $\Gamma_0(p)$ is zero. This will be done with the aid of the discriminant $\Delta = g_2{}^3 - 27g_3{}^2$.

We recall that $\Delta(\tau)$ is periodic with period 1 and has the Fourier expansion (Theorem 1.19)

$$\Delta(\tau) = (2\pi)^{12} \sum_{n=1}^{\infty} \tau(n)e^{2\pi i n\tau}$$

where the $\tau(n)$ are integers with $\tau(1) = 1$ and $\tau(2) = -24$. However, $\Delta(\tau)$ is not invariant under all transformations of Γ. In fact we have

$$\Delta\left(\frac{a\tau + b}{c\tau + d}\right) = (c\tau + d)^{12}\Delta(\tau) \quad \text{if} \quad \begin{pmatrix} a & b \\ c & d \end{pmatrix} \in \Gamma.$$

In particular,

$$\Delta(\tau + 1) = \Delta(\tau) \quad \text{and} \quad \Delta\left(\frac{-1}{\tau}\right) = \tau^{12}\Delta(\tau).$$

Even though $\Delta(\tau)$ is not invariant under Γ it can be used to construct functions automorphic under the subgroup $\Gamma_0(q)$ for each integer q.

Theorem 4.7. *For a fixed integer q, let*

$$\varphi(\tau) = \frac{\Delta(q\tau)}{\Delta(\tau)} \quad \text{if } \tau \in H.$$

Then φ is automorphic under $\Gamma_0(q)$. Moreover, the Fourier expansion of φ has the form

$$\varphi(\tau) = x^{q-1}\left(1 + \sum_{n=1}^{\infty} b_n x^n\right),$$

where the b_n are integers and $x = e^{2\pi i\tau}$.

PROOF. First we obtain the Fourier expansion. We have

$$\Delta(\tau) = (2\pi)^{12} \sum_{n=1}^{\infty} \tau(n)x^n = (2\pi)^{12}x\left\{1 + \sum_{n=1}^{\infty} \tau(n + 1)x^n\right\},$$

where $x = e^{2\pi i\tau}$. Hence

$$\Delta(q\tau) = (2\pi)^{12}x^q\left\{1 + \sum_{n=1}^{\infty} \tau(n + 1)x^{nq}\right\}$$

so

$$\varphi(\tau) = \frac{\Delta(q\tau)}{\Delta(\tau)} = x^{q-1}\frac{1 + \sum_{n=1}^{\infty} \tau(n + 1)x^{nq}}{1 + \sum_{n=1}^{\infty} \tau(n + 1)x^n} = x^{q-1}\left(1 + \sum_{n=1}^{\infty} b_n x^n\right)$$

where the b_n are integers.

Now φ is clearly meromorphic in H, and we will prove next that φ is invariant under $\Gamma_0(q)$.

If $V = \begin{pmatrix} a & b \\ c & d \end{pmatrix} \in \Gamma_0(q)$ then $c = c_1 q$ for some integer c_1. Hence

$$\Delta(V\tau) = (c\tau + d)^{12}\Delta(\tau) = (c_1 q\tau + d)^{12}\Delta(\tau).$$

On the other hand,

$$qV\tau = q\frac{a\tau + b}{c\tau + d} = \frac{a(q\tau) + bq}{c_1(q\tau) + d} = W(q\tau),$$

where

$$W = \begin{pmatrix} a & bq \\ c_1 & d \end{pmatrix}.$$

But $W \in \Gamma$ because $\det W = ad - bc_1 q = ad - bc = 1$. Hence

$$\Delta(qV\tau) = \Delta(W(q\tau)) = (c_1(q\tau) + d)^{12}\Delta(q\tau),$$

so

$$\varphi(V\tau) = \frac{\Delta(qV\tau)}{\Delta(V\tau)} = \frac{(c_1 q\tau + d)^{12}\Delta(q\tau)}{(c_1 q\tau + d)^{12}\Delta(\tau)} = \varphi(\tau).$$

This completes the proof. □

Now φ has a zero of order $q - 1$ at ∞ and no further zeros in H. Next we show that φ does not vanish at the vertex $\tau = 0$ of the fundamental region of $\Gamma_0(q)$. In fact, we show that $\varphi(\tau) \to \infty$ as $\tau \to 0$.

Theorem 4.8. *If $\tau \in H$ we have*

$$\varphi\left(\frac{-1}{q\tau}\right) = \frac{1}{q^{12}\varphi(\tau)}.$$

Hence $\varphi(\tau) \to \infty$ as $\tau \to 0$.

PROOF. Since $\Delta(-1/\tau) = \tau^{12}\Delta(\tau)$ we have

$$\Delta\left(-\frac{1}{q\tau}\right) = (q\tau)^{12}\Delta(q\tau)$$

so

$$\varphi\left(\frac{-1}{q\tau}\right) = \frac{\Delta\left(q\frac{-1}{q\tau}\right)}{\Delta\left(\frac{-1}{q\tau}\right)} = \frac{\Delta\left(-\frac{1}{\tau}\right)}{\Delta\left(-\frac{1}{q\tau}\right)} = \frac{\tau^{12}\Delta(\tau)}{(q\tau)^{12}\Delta(q\tau)} = q^{-12}\frac{1}{\varphi(\tau)}.$$

Since φ has a zero at ∞ we have $\varphi(-1/(q\tau)) \to 0$ as $\tau \to 0$ so $\varphi(\tau) \to \infty$. $\qquad\square$

4.8 The univalent function $\Phi(\tau)$

The function φ has a zero of order $q - 1$ at ∞ and no further zeros so its valence is $q - 1$. We seek a *univalent* function automorphic under $\Gamma_0(q)$ and this suggests that we consider φ^α, where $\alpha = 1/(q - 1)$. The Fourier expansion of φ^α need not have integer coefficients, since

$$\varphi^\alpha(\tau) = x\left(1 + \sum_{n=1}^\infty b_n x^n\right)^\alpha.$$

On the other hand we have the product representation

$$\Delta(\tau) = (2\pi)^{12} x \prod_{n=1}^\infty (1 - x^n)^{24}$$

so

$$\varphi(\tau) = \frac{\Delta(q\tau)}{\Delta(\tau)} = x^{q-1}\frac{\prod_{n=1}^\infty (1 - x^{qn})^{24}}{\prod_{n=1}^\infty (1 - x^n)^{24}}$$

$$= x^{q-1}\left(1 + \sum_{n=1}^\infty d_q(n)x^n\right)^{24}$$

where the coefficients $d_q(n)$ are integers. Therefore if $\alpha = 1/(q - 1)$ we have

(5) $$\varphi^\alpha(\tau) = x\left(1 + \sum_{n=1}^\infty d_q(n)x^n\right)^{24\alpha}$$

and the Fourier series for $\varphi^\alpha(\tau)$ will certainly have integer coefficients if 24α is an integer, that is, if $q - 1$ divides 24. This occurs when $q = 2, 3, 4, 5, 7, 9, 13,$ and 25.

Definition. If $q - 1$ divides 24 let $\alpha = 1/(q - 1)$ and $r = 24\alpha$. We define the function Φ by the relations

$$\Phi(\tau) = \varphi^\alpha(\tau) = \left(\frac{\Delta(q\tau)}{\Delta(\tau)}\right)^\alpha = \left(\frac{\eta(q\tau)}{\eta(\tau)}\right)^r.$$

The function Φ so defined is analytic and nonzero in H. The Fourier series for Φ in (5) shows that Φ has a first order zero at ∞ and that

$$\frac{1}{\Phi(\tau)} = \frac{1}{x} + I(x),$$

where $I(x)$ is a power series in x with integer coefficients.

Since φ is automorphic under $\Gamma_0(q)$ we have $\varphi(V\tau) = \varphi(\tau)$ for every element V of $\Gamma_0(q)$. Hence, extracting roots of order $q - 1$, we have

$$\Phi(V\tau) = \varepsilon\Phi(\tau)$$

where $\varepsilon^{q-1} = 1$. The next theorem shows that, in fact, $\varepsilon = 1$ whenever $24/(q - 1)$ is an even integer and q is prime. This occurs when $q = 2, 3, 5, 7$, and 13. For these values of q the function Φ is automorphic under $\Gamma_0(q)$.

4.9 Invariance of $\Phi(\tau)$ under transformations of $\Gamma_0(q)$

The properties of Dedekind sums proved in the foregoing chapter lead to a simple proof of the invariance of the univalent function $\Phi(\tau)$.

Theorem 4.9. *Let $q = 2, 3, 5, 7$, or 13, and let $r = 24/(q - 1)$. Then the function*

$$(6) \qquad \Phi(\tau) = \left(\frac{\eta(q\tau)}{\eta(\tau)}\right)^r$$

is automorphic under the subgroup $\Gamma_0(q)$.

PROOF. If $q = 2$ we have $r = 24$ and $\Phi(\tau) = \Delta(q\tau)/\Delta(\tau)$. In this case the theorem was already proved in Theorem 4.7. Therefore we shall assume that $q \geq 3$.

Let $V = \begin{pmatrix} a & b \\ c & d \end{pmatrix}$ be any element of $\Gamma_0(q)$. Then $ad - bc = 1$ and $c \equiv 0 \pmod{q}$. We can suppose that $c \geq 0$. If $c = 0$ then V is a power of the translation $T\tau = \tau + 1$, and since $\eta(\tau + 1) = e^{\pi i/12}\eta(\tau)$ we find

$$\Phi(\tau + 1) = \left(\frac{\eta(q\tau + q)}{\eta(\tau + 1)}\right)^r = e^{\pi i r(q-1)/12}\Phi(\tau) = \Phi(\tau).$$

Therefore we can assume that $c > 0$ and that $c = c_1 q$, where $c_1 > 0$. Dedekind's functional equation for $\eta(\tau)$ gives us

$$(7) \qquad \eta(V\tau) = \varepsilon(V)\{-i(c\tau + d)\}^{1/2}\eta(\tau)$$

where

$$(8) \qquad \varepsilon(V) = \exp\left\{\pi i\left(\frac{a + d}{12c} + s(-d, c)\right)\right\}.$$

87

We also have

$$\eta(qV\tau) = \eta\left(\frac{a(q\tau) + bq}{c_1(q\tau) + d}\right) = \eta(V_1q\tau)$$

where

$$V_1 = \begin{pmatrix} a & bq \\ c_1 & d \end{pmatrix}.$$

Since $V_1 \in \Gamma$ we have

$$\eta(qV\tau) = \varepsilon(V_1)\{-i(c_1q\tau + d)\}^{1/2}\eta(q\tau)$$

which, together with (7), gives us

$$\Phi(V\tau) = \left(\frac{\varepsilon(V_1)}{\varepsilon(V)}\right)^r \Phi(\tau).$$

But (8) shows that $(\varepsilon(V_1)/\varepsilon(V))^r = e^{-\pi i r\delta}$, where

$$\delta = \left\{\frac{a+d}{12c} + s(-d, c)\right\} - \left\{\frac{a+d}{12c_1} + s(-d, c_1)\right\}.$$

Since $ad - bc = 1$ we have $ad \equiv 1 \pmod{c}$ and $ad \equiv 1 \pmod{c_1}$ so $s(-d, c) = -s(a, c)$ and $s(-d, c_1) = -s(a, c_1)$, and Theorem 3.11 shows that $r\delta$ is an even integer. Therefore $e^{-\pi i r\delta} = 1$ and $\Phi(V\tau) = \Phi(\tau)$. □

4.10 The function j_p expressed as a polynomial in Φ

If p is prime and if f is automorphic under Γ, we have shown that the function

$$f_p(\tau) = \frac{1}{p}\sum_{\lambda=0}^{p-1} f\left(\frac{\tau + \lambda}{p}\right)$$

is automorphic under $\Gamma_0(p)$, and its Fourier coefficients consist of every pth coefficient of f. To obtain divisibility properties of the coefficients of $j_p(\tau)$ we shall express j_p as a polynomial in the function Φ.

In deriving the differential equation for the Weierstrass $\wp$ function we formed a linear combination of $\wp$, $\wp^2$ and $\wp^3$ which gave a principal part near $z = 0$ equal to that of $[\wp'(z)]^2$. The procedure here is analogous. Both functions j_p and Φ have a pole at the vertex $\tau = 0$ of the fundamental region of $\Gamma_0(p)$. We form a linear combination of powers of Φ to obtain a principal part equal to that of j_p.

To obtain the order of the pole of $j_p(\tau)$ at $\tau = 0$ we use Theorem 4.6 which gives us the relation

$$j_p\left(-\frac{1}{\tau}\right) = j_p(\tau) + \frac{1}{p}\,j(p\tau) - \frac{1}{p}\,j\left(\frac{\tau}{p}\right)$$

valid for prime p. Replacing τ by $p\tau$ in this formula we obtain

Theorem 4.10. *If p is prime and $\tau \in H$ then*

$$j_p\left(-\frac{1}{p\tau}\right) = j_p(p\tau) + \frac{1}{p}\,j(p^2\tau) - \frac{1}{p}\,j(\tau).$$

Hence if $x = e^{2\pi i\tau}$ we have the Fourier expansion

$$p\,j_p\left(-\frac{1}{p\tau}\right) = x^{-p^2} - x^{-1} + I(x),$$

where $I(x)$ is a power series in x with integer coefficients.

PROOF. We have

$$j(\tau) = x^{-1} + c(0) + c(1)x + c(2)x^2 + \cdots,$$
$$j_p(\tau) = c(0) + c(p)x + c(2p)x^2 + \cdots,$$
$$p\,j_p(p\tau) = pc(0) + pc(p)x^p + pc(2p)x^{2p} + \cdots,$$

and

$$j(p^2\tau) = x^{-p^2} + c(0) + c(1)x^{p^2} + c(2)x^{2p^2} + \cdots,$$

so

$$p\,j_p\left(-\frac{1}{p\tau}\right) = p\,j_p(p\tau) + j(p^2\tau) - j(\tau)$$

$$= x^{-p^2} - x^{-1} + I(x). \qquad \square$$

Now we can express j_p as a polynomial in Φ.

Theorem 4.11. *Assume $p = 2, 3, 5, 7$ or 13, and let*

$$\Phi(\tau) = \left(\frac{\eta(p\tau)}{\eta(\tau)}\right)^r, \quad \text{where } r = \frac{24}{p-1}.$$

Then there exist integers $a_1, \ldots, a_{p^2}$ such that

$$(9) \qquad j_p(\tau) = p^{r/2-1}\{a_1\Phi(\tau) + a_2\Phi^2(\tau) + \cdots + a_{p^2}\Phi^{p^2}(\tau)\} + c(0).$$

PROOF. By Theorem 4.10 we have

$$p\,j_p\left(-\frac{1}{p\tau}\right) = x^{-p^2} - x^{-1} + I(x),$$

and, since $12\alpha = r/2$, Theorem 4.8 gives us

$$p^{r/2}\Phi\left(-\frac{1}{p\tau}\right) = \frac{1}{\Phi(\tau)} = x^{-1} + I(x).$$

Let $\psi(\tau) = p^{r/2}\Phi(-1/(p\tau))$. Then the difference

$$pj_p\left(-\frac{1}{p\tau}\right) - \{\psi(\tau)\}^{p^2}$$

has a pole of order $\leq p^2 - 1$ at $x = 0$, and the Laurent expansion near $x = 0$ has integer coefficients. Hence there is an integer b_1 such that

$$pj_p\left(-\frac{1}{p\tau}\right) - \{\psi(\tau)\}^{p^2} - b_1\{\psi(\tau)\}^{p^2-1}$$

has a pole of order $\leq p^2 - 2$ at $x = 0$, and the Laurent expansion near $x = 0$ has integer coefficients. In p^2 steps we arrive at a function

$$f\left(-\frac{1}{p\tau}\right) = pj_p\left(-\frac{1}{p\tau}\right) - \{\psi(\tau)\}^{p^2} - b_1\{\psi(\tau)\}^{p^2-1} - \cdots - b_{p^2-1}\psi(\tau)$$

which is *analytic* at $x = 0$ and has a power series expansion with integer coefficients. Moreover, all the numbers $b_1, \ldots, b_{p^2-1}$ are integers. Replacing τ by $-1/(p\tau)$ we obtain

$$f(\tau) = pj_p(\tau) - \{p^{r/2}\Phi(\tau)\}^{p^2} - b_1\{p^{r/2}\Phi(\tau)\}^{p^2-1} - \cdots - b_{p^2-1}\{p^{r/2}\Phi(\tau)\}.$$

Now $f(\tau)$ is automorphic under $\Gamma_0(p)$ and analytic at each point τ in H. The function f is also analytic at the vertex $\tau = 0$ (by construction). Therefore f is bounded in H so f is constant. But this constant is $pc(0)$ since $\Phi(\tau)$ vanishes at ∞. Thus we find

$$pj_p(\tau) = \{p^{r/2}\Phi(\tau)\}^{p^2} + b_1\{p^{r/2}\Phi(\tau)\}^{p^2-1} + \cdots + b_{p^2-1}\{p^{r/2}\Phi(\tau)\} + pc(0)$$

so $j_p(\tau)$ is expressible as indicated in (9). $\qquad\square$

Theorem 4.12. *The coefficients in the Fourier expansion of $j(\tau)$ satisfy the following congruences:*

$$c(2n) \equiv 0 \pmod{2^{11}}$$
$$c(3n) \equiv 0 \pmod{3^5}$$
$$c(5n) \equiv 0 \pmod{5^2}$$
$$c(7n) \equiv 0 \pmod{7}.$$

PROOF. The previous theorem shows that for $p = 2, 3, 5, 7$ and 13 we have

$$c(pn) \equiv 0 \pmod{p^{(r/2)-1}},$$

where $r = 24/(p-1)$. Therefore we simply compute $(r/2) - 1$ to obtain the stated congruences. Note that $(r/2) - 1 = 0$ when $p = 13$ so we get a trivial congruence in this case. $\qquad\square$

Note. By repeated application of the foregoing ideas Lehner [24] derived the following more general congruences, valid for $\alpha \geq 1$:

$$c(2^\alpha n) \equiv 0 \pmod{2^{3\alpha + 8}}$$
$$c(3^\alpha n) \equiv 0 \pmod{3^{2\alpha + 3}}$$
$$c(5^\alpha n) \equiv 0 \pmod{5^{\alpha + 1}}$$
$$c(7^\alpha n) \equiv 0 \pmod{7^\alpha}.$$

Since it is known that $c(13)$ is not divisible by 13, congruences of the above type cannot exist for 13. In 1958 Morris Newman [30] found congruences of a different kind for 13. He showed that

$$c(13np) + c(13n)c(13p) + p^{-1}c\left(\frac{13n}{p}\right) \equiv 0 \pmod{13},$$

where $p^{-1}p \equiv 1 \pmod{13}$ and $c(x) = 0$ if x is not an integer. The congruences of Lehner and Newman were generalized by Atkin and O'Brien [5] in 1967.

Exercises for Chapter 4

1. This exercise relates the Dedekind function $\eta(\tau)$ to the Jacobi theta function $\vartheta(\tau)$ defined on H by the equation

$$\vartheta(\tau) = 1 + 2\sum_{n=1}^{\infty} e^{\pi i n^2 \tau} = \sum_{n=-\infty}^{\infty} e^{\pi i n^2 \tau}.$$

The definition shows that ϑ is analytic in H and periodic with period 2.
   Jacobi's triple product identity (Theorem 14.6 in [4]) states that

$$\prod_{n=1}^{\infty}(1 - x^{2n})(1 + x^{2n-1}z^2)(1 + x^{2n-1}z^{-2}) = \sum_{m=-\infty}^{\infty} x^{m^2}z^{2m}$$

if $z \neq 0$ and $|x| < 1$.
   (a) Show that x and z can be chosen to give the product representation

$$\vartheta(\tau) = \prod_{n=1}^{\infty}(1 - e^{2\pi i n\tau})(1 + e^{(2n-1)\pi i\tau})^2.$$

This implies that $\vartheta(\tau)$ is never zero in H.
   (b) If $\tau \in H$ prove that

$$\vartheta(\tau) = \frac{\eta^2\left(\dfrac{\tau + 1}{2}\right)}{\eta(\tau + 1)}.$$

   (c) Prove that $\vartheta(-1/\tau) = (-i\tau)^{1/2}\vartheta(\tau)$.
      Hint: If $S\tau = -1/\tau$, find elements A and B of Γ such that

$$\frac{S\tau + 1}{2} = A\left(\frac{\tau + 1}{2}\right) \qquad \text{and} \qquad S\tau + 1 = B(\tau + 1).$$

2. Let G denote the subgroup of Γ generated by the transformations S and T^2, where $S\tau = -1/\tau$ and $T\tau = \tau + 1$.

(a) If $\begin{pmatrix} a & b \\ c & d \end{pmatrix} \in G$ prove that $a \equiv d \pmod 2$ and $b \equiv c \pmod 2$.

(b) If $V \in G$ prove that there exist elements A and B of Γ such that

$$\frac{V\tau + 1}{2} = A\left(\frac{\tau + 1}{2}\right) \quad \text{and} \quad V\tau + 1 = B(\tau + 1).$$

(c) If $\begin{pmatrix} a & b \\ c & d \end{pmatrix} \in G$ and $c > 0$ prove that

$$\vartheta\left(\frac{a\tau + b}{c\tau + d}\right) = \varepsilon(a, b, c, d)\{-i(c\tau + d)\}^{1/2}\vartheta(\tau),$$

where $|\varepsilon(a, b, c, d)| = 1$. Express $\varepsilon(a, b, c, d)$ in terms of Dedekind sums.

Exercises 3 through 8 outline a proof (due to Mordell [28]) of the multiplicativity of Ramanujan's function $\tau(n)$. We recall that

$$\sum_{n=1}^{\infty} \tau(n)e^{2\pi i n\tau} = (2\pi)^{-12}\Delta(\tau) = e^{2\pi i\tau} \prod_{m=1}^{\infty} (1 - e^{2\pi i m\tau})^{24}.$$

3. Let p be a prime and let k be an integer, $1 \le k \le p - 1$. Show that there exists an integer h such that

$$\tau^{12}\Delta\left(\frac{\tau + h}{p}\right) = \Delta\left(\frac{k\tau - 1}{p\tau}\right)$$

and that h runs through a reduced residue system mod p with k.

4. If p is a prime, define

$$F_p(\tau) = p^{11}\Delta(p\tau) + \frac{1}{p}\sum_{k=0}^{p-1} \Delta\left(\frac{\tau + k}{p}\right).$$

Prove that:

(a) $F_p(\tau + 1) = F_p(\tau)$;   (b) $F_p\left(\dfrac{-1}{\tau}\right) = \tau^{12}F_p(\tau)$.

Note: Exercise 3 will be helpful for part (b).

5. Prove that $F_p(\tau) = \tau(p)\Delta(\tau)$, where $\tau(p)$ is Ramanujan's function.

6. Use Exercises 4 and 5 to deduce the formulas

(a) $\tau(p^{n+1}) = \tau(p)\tau(p^n) - p^{11}\tau(p^{n-1})$ for $n \ge 1$.

(b) $\tau(p^\alpha n) = \tau(p)\tau(p^{\alpha-1}n) - p^{11}\tau(p^{\alpha-2}n)$ for $\alpha \ge 2$ and $(n, p) = 1$.

7. If α is an integer, $\alpha \ge 0$, and if $(n, p) = 1$, let

$$g(\alpha) = \tau(p^\alpha n) - \tau(p^\alpha)\tau(n).$$

Show that $g(\alpha + 1)$ is a linear combination of $g(\alpha)$ and $g(\alpha - 1)$ for $\alpha \ge 2$ and deduce that $g(\alpha) = 0$ for all α.

8. Prove that

$$\tau(m)\tau(n) = \sum_{d|(m,n)} d^{11}\tau\left(\frac{mn}{d^2}\right).$$

In particular, when $(m, n) = 1$ this implies $\tau(m)\tau(n) = \tau(mn)$.

9. If $\tau \in H$ and $x = e^{2\pi i \tau}$ prove that

$$\left\{504 \sum_{n=0}^{\infty} \sigma_5(n)x^n\right\}^2 = \{j(\tau) - 12^3\} \sum_{n=1}^{\infty} \tau(n)x^n,$$

where $\sigma_5(0) = -1/504$. Equate coefficients of x^n to obtain the identity

$$(504)^2 \sum_{k=0}^{n} \sigma_5(k)\sigma_5(n - k) = \tau(n + 1) - 984\tau(n) + \sum_{k=1}^{n-1} c(k)\tau(n - k).$$

10. Use Exercise 9 together with Exercise 10 of Chapter 6 to prove that

$$\frac{65520}{691} \{\sigma_{11}(n) - \tau(n)\} = \tau(n + 1) + 24\tau(n) + \sum_{k=1}^{n-1} c(k)\tau(n - k).$$

This formula, due to Lehmer [20], can be used to determine the coefficients $c(n)$ recursively in terms of $\tau(n)$. Since the right member is an integer, the formula also implies Ramanujan's remarkable congruence

$$\tau(n) \equiv \sigma_{11}(n) \pmod{691}.$$

5 Rademacher's series for the partition function

5.1 Introduction

The unrestricted partition function $p(n)$ counts the number of ways a positive integer n can be expressed as a sum of positive integers $\leq n$. The number of summands is unrestricted, repetition is allowed, and the order of the summands is not taken into account.

The partition function is generated by Euler's infinite product

$$(1) \qquad F(x) = \prod_{m=1}^{\infty} \frac{1}{1 - x^m} = \sum_{n=0}^{\infty} p(n)x^n,$$

where $p(0) = 1$. Both the product and series converge absolutely and represent the analytic function F in the unit disk $|x| < 1$. A proof of (1) and other elementary properties of $p(n)$ can be found in Chapter 14 of [4]. This chapter is concerned with the behavior of $p(n)$ for large n.

The partition function $p(n)$ satisfies the asymptotic relation

$$p(n) \sim \frac{e^{K\sqrt{n}}}{4n\sqrt{3}} \quad \text{as } n \to \infty,$$

where $K = \pi(2/3)^{1/2}$. This was first discovered by Hardy and Ramanujan [13] in 1918 and, independently, by J. V. Uspensky [52] in 1920. Hardy and Ramanujan proved more. They obtained a remarkable asymptotic formula of the form

$$(2) \qquad p(n) = \sum_{k < \alpha\sqrt{n}} P_k(n) + O(n^{-1/4}),$$

where α is a constant and $P_1(n)$ is the dominant term, asymptotic to $e^{K\sqrt{n}}/(4n\sqrt{3})$. The terms $P_2(n)$, $P_3(n)$, ... are of similar type but with smaller constants in place of K in the exponential. Since $p(n)$ is an integer the finite sum on the right of (2) gives $p(n)$ exactly when n is large enough to insure that the error term is less than $1/2$. This is a rare example of a formula which is both asymptotic and exact. As is often the case with asymptotic formulas of this type, the infinite sum

$$(3) \qquad\qquad \sum_{k=1}^{\infty} P_k(n)$$

diverges for each n. The divergence of (3) was shown by D. H. Lehmer [21] in 1937.

Hans Rademacher, while preparing lecture notes in 1937 on the work of Hardy and Ramanujan, made a small change in the analysis which resulted in slightly different terms $R_k(n)$ in place of the $P_k(n)$ in (2). This had a profound effect on the final result since, instead of (2), Rademacher obtained a *convergent* series,

$$(4) \qquad\qquad p(n) = \sum_{k=1}^{\infty} R_k(n).$$

The exact form of the Rademacher terms $R_k(n)$ is described below in Theorem 5.10. Rademacher [35] also showed that the remainder after N terms is $O(n^{-1/4})$ when N is of order $\sqrt{n}$, in agreement with (2).

This chapter is devoted to a proof of Rademacher's exact formula for $p(n)$. The proof is of special interest because it represents one of the crowning achievements of the so-called "circle method" of Hardy, Ramanujan and Littlewood which has been highly successful in many asymptotic problems of additive number theory. The proof also displays a marvelous application of Dedekind's modular function $\eta(\tau)$.

5.2 The plan of the proof

This section gives a rough sketch of the proof. The starting point is Euler's formula (1) which implies

$$\frac{F(x)}{x^{n+1}} = \sum_{k=0}^{\infty} \frac{p(k)x^k}{x^{n+1}} \quad \text{if } 0 < |x| < 1,$$

for each $n \geq 0$. The last series is the Laurent expansion of $F(x)/x^{n+1}$ in the punctured disk $0 < |x| < 1$. This function has a pole at $x = 0$ with residue $p(n)$ so by Cauchy's residue theorem we have

$$p(n) = \frac{1}{2\pi i} \int_C \frac{F(x)}{x^{n+1}} \, dx,$$

where C is any positively oriented simple closed contour which lies inside the unit circle and encloses the origin. The basic idea of the circle method is to choose a contour C which lies near the singularities of the function $F(x)$.

The factors in the product defining $F(x)$ vanish whenever $x = 1$, $x^2 = 1$, $x^3 = 1$, etc., so each root of unity is a singularity of $F(x)$. The circle method chooses a circular contour C of radius nearly 1 and divides C into arcs $C_{h,k}$ lying near the roots of unity $e^{2\pi i h/k}$, where $0 \le h < k$, $(h, k) = 1$, and $k = 1, 2, \ldots, N$. The integral along C can be written as a finite sum of integrals along these arcs,

$$\int_C = \sum_{k=1}^{N} \sum_{\substack{h=0 \\ (h,k)=1}}^{k-1} \int_{C_{h,k}}.$$

On each arc $C_{h,k}$ the function $F(x)$ in the integrand is replaced by an elementary function $\psi_{h,k}(x)$ which has essentially the same behavior as F near the singularity $e^{2\pi i h/k}$. This elementary function $\psi_{h,k}$ arises naturally from the functional equation satisfied by the Dedekind eta function $\eta(\tau)$. The functions F and η are related by the equation

$$F(e^{2\pi i \tau}) = e^{\pi i \tau/12}/\eta(\tau),$$

and the functional equation for η gives a formula which describes the behavior of F near each singularity $e^{2\pi i h/k}$. The replacement of F by $\psi_{h,k}$ introduces an error which needs to be estimated. The integrals of the $\psi_{h,k}$ along $C_{h,k}$ are then evaluated, and their sum over h produces the term $R_k(n)$ in Rademacher's series.

In 1943 Rademacher [38] modified the circle method by replacing the circular contour C by another contour in the τ-plane, where $x = e^{2\pi i \tau}$. This new path of integration simplifies the estimates that need to be made and clarifies the manner in which the singularities contribute to the final formula.

The next section expresses Dedekind's functional equation in terms of F. Sections 5.5 and 5.6 describe the path of integration used by Rademacher, and Section 5.7 carries out the plan outlined above.

5.3 Dedekind's functional equation expressed in terms of F

Theorem 5.1. *Let* $F(t) = 1/\prod_{m=1}^{\infty} (1 - t^m)$ *and let*

$$(5) \qquad x = \exp\left(\frac{2\pi i h}{k} - \frac{2\pi z}{k^2}\right), \qquad x' = \exp\left(\frac{2\pi i H}{k} - \frac{2\pi}{z}\right),$$

where $\mathrm{Re}(z) > 0$, $k > 0$, $(h, k) = 1$, *and* $hH \equiv -1 \pmod{k}$. *Then*

$$(6) \qquad F(x) = e^{\pi i s(h,k)}\left(\frac{z}{k}\right)^{1/2} \exp\left(\frac{\pi}{12z} - \frac{\pi z}{12k^2}\right) F(x').$$

Note. If $|z|$ is small, the point x in (5) lies near the root of unity $e^{2\pi i h/k}$, whereas x' lies near the origin. Hence $F(x')$ is nearly $F(0) = 1$, and Equation (6) gives the behavior of F near the singularity $e^{2\pi i h/k}$. Aside from a constant factor, for small $|z|$, F behaves like

$$z^{1/2} \exp\left(\frac{\pi}{12z}\right).$$

PROOF. If $\begin{pmatrix} a & b \\ c & d \end{pmatrix} \in \Gamma$ with $c > 0$, the functional equation for $\eta(\tau)$ implies

$$(7) \qquad \frac{1}{\eta(\tau)} = \frac{1}{\eta(\tau')}\{-i(c\tau + d)\}^{1/2} \exp\left\{\pi i\left(\frac{a + d}{12c} + s(-d, c)\right)\right\},$$

where $\tau' = (a\tau + b)/(c\tau + d)$. Since $F(e^{2\pi i\tau}) = e^{\pi i\tau/12}/\eta(\tau)$, (7) implies

$$(8) \qquad F(e^{2\pi i\tau}) = F(e^{2\pi i\tau'}) \exp\left(\frac{\pi i(\tau - \tau')}{12}\right)\{-i(c\tau + d)\}^{1/2}$$

$$\times \exp\left\{\pi i\left(\frac{a + d}{12c} + s(-d, c)\right)\right\}.$$

Now choose

$$a = H, c = k, d = -h, b = -\frac{hH + 1}{k}, \qquad \text{and} \qquad \tau = \frac{iz + h}{k}.$$

Then

$$\tau' = \frac{iz^{-1} + H}{k}$$

and (8) becomes

$$F\left(\exp\left(\frac{2\pi i h}{k} - \frac{2\pi z}{k}\right)\right) = F\left(\exp\left(\frac{2\pi i H}{k} - \frac{2\pi}{kz}\right)\right)z^{1/2}$$

$$\times \exp\left\{\frac{\pi}{12kz} - \frac{\pi z}{12k} + \pi i s(h, k)\right\}.$$

When z is replaced by z/k this gives (6). $\qquad \square$

5.4 Farey fractions

Our next task is to describe the path of integration used by Rademacher. The path is related to a set of reduced fractions in the unit interval called *Farey fractions*. This section describes these fractions and some of their properties.

Definition. The set of Farey fractions of order n, denoted by F_n, is the set of reduced fractions in the closed interval $[0, 1]$ with denominators $\leq n$, listed in increasing order of magnitude.

EXAMPLES

$F_1: \frac{0}{1}, \frac{1}{1}$

$F_2: \frac{0}{1}, \frac{1}{2}, \frac{1}{1}$

$F_3: \frac{0}{1}, \frac{1}{3}, \frac{1}{2}, \frac{2}{3}, \frac{1}{1}$

$F_4: \frac{0}{1}, \frac{1}{4}, \frac{1}{3}, \frac{1}{2}, \frac{2}{3}, \frac{3}{4}, \frac{1}{1}$

$F_5: \frac{0}{1}, \frac{1}{5}, \frac{1}{4}, \frac{1}{3}, \frac{2}{5}, \frac{1}{2}, \frac{3}{5}, \frac{2}{3}, \frac{3}{4}, \frac{4}{5}, \frac{1}{1}$

$F_6: \frac{0}{1}, \frac{1}{6}, \frac{1}{5}, \frac{1}{4}, \frac{1}{3}, \frac{2}{5}, \frac{1}{2}, \frac{3}{5}, \frac{2}{3}, \frac{3}{4}, \frac{4}{5}, \frac{5}{6}, \frac{1}{1}$

$F_7: \frac{0}{1}, \frac{1}{7}, \frac{1}{6}, \frac{1}{5}, \frac{1}{4}, \frac{2}{7}, \frac{1}{3}, \frac{2}{5}, \frac{3}{7}, \frac{1}{2}, \frac{4}{7}, \frac{3}{5}, \frac{2}{3}, \frac{5}{7}, \frac{3}{4}, \frac{4}{5}, \frac{5}{6}, \frac{6}{7}, \frac{1}{1}$

These examples illustrate some general properties of Farey fractions. For example, $F_n \subset F_{n+1}$, so we get F_{n+1} by inserting new fractions in F_n. If $(a/b) < (c/d)$ are consecutive in F_n and separated in F_{n+1}, then the fraction $(a + c)/(b + d)$ does the separating, and no new ones are inserted between a/b and c/d. This new fraction is called the *mediant* of a/b and c/d.

Theorem 5.2. *If $(a/b) < (c/d)$, their mediant $(a + c)/(b + d)$ lies between them.*

PROOF

$$\frac{a + c}{b + d} - \frac{a}{b} = \frac{bc - ad}{b(b + d)} > 0 \quad \text{and} \quad \frac{c}{d} - \frac{a + c}{b + d} = \frac{bc - ad}{d(b + d)} > 0. \quad \square$$

The above examples show that $\frac{1}{3}$ and $\frac{2}{3}$ are consecutive fractions in F_n for $n = 5, 6,$ and 7. This illustrates the following general property.

Theorem 5.3. *Given $0 \leq a/b < c/d \leq 1$. If $bc - ad = 1$ then a/b and c/d are consecutive terms in F_n for the following values of n:*

$$\max(b, d) \leq n \leq b + d - 1.$$

PROOF. The condition $bc - ad = 1$ implies that a/b and c/d are in lowest terms. If $\max(b, d) \leq n$ then $b \leq n$ and $d \leq n$ so a/b and c/d are certainly in F_n. Now we prove they are *consecutive* if $n \leq b + d - 1$. If they are *not* consecutive there is another fraction h/k between them, $a/b < h/k < c/d$. But now we can show that $k \geq b + d$ because we have the identity

(9) $$k = (bc - ad)k = b(ck - dh) + d(bh - ak).$$

But the inequalities $a/b < h/k < c/d$ show that $ck - dh \geq 1$ and $bh - ak \geq 1$ so $k \geq b + d$. Thus, any fraction h/k that lies between a/b and c/d has denominator $k \geq b + d$. Therefore, if $n \leq b + d - 1$, then a/b and c/d must be consecutive in F_n. This completes the proof. $\square$

Equation (9) also yields the following theorem.

Theorem 5.4. *Given $0 \le a/b < c/d \le 1$ with $bc - ad = 1$, let h/k be the mediant of a/b and c/d. Then $a/b < h/k < c/d$, and these fractions satisfy the unimodular relations*

$$bh - ak = 1, \qquad ck - dh = 1.$$

PROOF. Since h/k lies between a/b and c/d we have $bh - ak \ge 1$ and $ck - dh \ge 1$. Equation (9) shows that $k = b + d$ if, and only if, $bh - ak = ck - dh = 1$. □

The foregoing theorems tell us how to construct F_{n+1} from F_n.

Theorem 5.5. *The set F_{n+1} includes F_n. Each fraction in F_{n+1} which is not in F_n is the mediant of a pair of consecutive fractions in F_n. Moreover, if $a/b < c/d$ are consecutive in any F_n, then they satisfy the unimodular relation $bc - ad = 1$.*

PROOF. We use induction on n. When $n = 1$ the fractions $0/1$ and $1/1$ are consecutive and satisfy the unimodular relation. We pass from F_1 to F_2 by inserting the mediant $1/2$. Now suppose a/b and c/d are consecutive in F_n and satisfy the unimodular relation $bc - ad = 1$. By Theorem 5.3, they will be consecutive in F_m for all m satisfying

$$\max(b, d) \le m \le b + d - 1.$$

Form their mediant h/k, where $h = a + c$, $k = b + d$. By Theorem 5.4 we have $bh - ak = 1$ and $ck - dh = 1$ so h and k are relatively prime. The fractions a/b and c/d are consecutive in F_m for all m satisfying $\max(b, d) \le m \le b + d - 1$, but are *not* consecutive in F_k since $k = b + d$ and h/k lies in F_k between a/b and c/d. But the two new pairs $a/b < h/k$ and $h/k < c/d$ are now consecutive in F_k because $k = \max(b, k)$ and $k = \max(d, k)$. The new consecutive pairs still satisfy the unimodular relations $bh - ak = 1$ and $ck - dh = 1$. This shows that in passing from F_n to F_{n+1} every new fraction inserted must be the mediant of a consecutive pair in F_n, and the new consecutive pairs satisfy the unimodular relations. Therefore F_{n+1} has these properties if F_n does. □

5.5 Ford circles

Definition. Given a rational number h/k with $(h, k) = 1$. The *Ford circle* belonging to this fraction is denoted by $C(h, k)$ and is that circle in the complex plane with radius $1/(2k^2)$ and center at the point $(h/k) + i/(2k^2)$ (see Figure 5.1).

Ford circles are named after L. R. Ford [9] who first studied their properties in 1938.

99

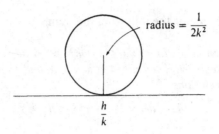

Figure 5.1   The Ford circle $C(h, k)$

Theorem 5.6. *Two Ford circles $C(a, b)$ and $C(c, d)$ are either tangent to each other or they do not intersect. They are tangent if, and only if, $bc - ad = \pm 1$. In particular, Ford circles of consecutive Farey fractions are tangent to each other.*

PROOF. The square of the distance D between centers is (see Figure 5.2)

$$D^2 = \left(\frac{a}{b} - \frac{c}{d}\right)^2 + \left(\frac{1}{2b^2} - \frac{1}{2d^2}\right)^2,$$

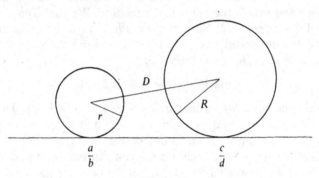

Figure 5.2

whereas the square of the sum of their radii is

$$(r + R)^2 = \left(\frac{1}{2b^2} + \frac{1}{2d^2}\right)^2.$$

The difference $D^2 - (r + R)^2$ is equal to

$$D^2 - (r + R)^2 = \left(\frac{ad - bc}{bd}\right)^2 + \left(\frac{1}{2b^2} - \frac{1}{2d^2}\right)^2 - \left(\frac{1}{2b^2} + \frac{1}{2d^2}\right)^2$$

$$= \frac{(ad - bc)^2 - 1}{b^2 d^2} \geq 0.$$

Moreover, equality holds if, and only if $(ad - bc)^2 = 1$.   □

Theorem 5.7. *Let $h_1/k_1 < h/k < h_2/k_2$ be three consecutive Farey fractions. The points of tangency of $C(h, k)$ with $C(h_1, k_1)$ and $C(h_2, k_2)$ are the points*

$$\alpha_1(h, k) = \frac{h}{k} - \frac{k_1}{k(k^2 + k_1{}^2)} + \frac{i}{k^2 + k_1{}^2}$$

and

$$\alpha_2(h, k) = \frac{h}{k} + \frac{k_2}{k(k^2 + k_2{}^2)} + \frac{i}{k^2 + k_2{}^2}.$$

Moreover, the point of contact $\alpha_1(h, k)$ lies on the semicircle whose diameter is the interval $[h_1/k_1, h/k]$.

PROOF. We refer to Figure 5.3. Write α_1 for $\alpha_1(h, k)$. The figure shows that

$$\alpha_1 = \left(\frac{h}{k} - a\right) + i\left(\frac{1}{2k^2} - b\right).$$

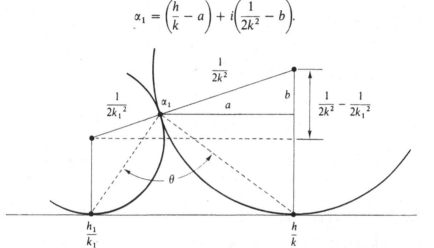

Figure 5.3

To determine a and b we refer to the similar right triangles and we get

$$\frac{a}{\dfrac{h}{k} - \dfrac{h_1}{k_1}} = \frac{\dfrac{1}{2k^2}}{\dfrac{1}{2k^2} + \dfrac{1}{2k_1{}^2}} = \frac{k_1{}^2}{k^2 + k_1{}^2}, \quad \text{so} \quad a = \frac{k_1}{k(k^2 + k_1{}^2)}.$$

Similarly, we find

$$\frac{b}{\dfrac{1}{2k^2}} = \frac{\dfrac{1}{2k^2} - \dfrac{1}{2k_1{}^2}}{\dfrac{1}{2k^2} + \dfrac{1}{2k_1{}^2}} = \frac{k_1{}^2 - k^2}{k_1{}^2 + k^2}, \quad \text{so} \quad b = \frac{1}{2k^2}\frac{k_1{}^2 - k^2}{k^2 + k_1{}^2}.$$

These give the required formula for α_1, and by analogy we get the corresponding formula for α_2.

To obtain the last statement, it suffices to show that the angle θ in Figure 5.3 is $\pi/2$. For this it suffices to show that the imaginary part of $\alpha_1(h, k)$ is the geometric mean of a and a', where

$$a = \frac{k_1}{k(k^2 + k_1{}^2)} \quad \text{and} \quad a' = \frac{h}{k} - \frac{h_1}{k_1} - a = \frac{1}{kk_1} - a.$$

(See Figure 5.4.) Now

$$aa' = \frac{k_1}{k(k^2 + k_1{}^2)} \left(\frac{1}{kk_1} - \frac{k_1}{k(k^2 + k_1{}^2)} \right)$$

$$= \frac{k_1}{k^2(k^2 + k_1{}^2)} \left(\frac{k^2}{k_1(k^2 + k_1{}^2)} \right) = \frac{1}{(k^2 + k_1{}^2)^2},$$

and this completes the proof. □

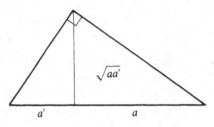

Figure 5.4

5.6 Rademacher's path of integration

For each integer N we construct a path $P(N)$ joining the points i and $i + 1$ as follows. Consider the Ford circles for the Farey series F_N. If $h_1/k_1 < h/k < h_2/k_2$ are consecutive in F_N, the points of tangency of $C(h_1, k_1)$, $C(h, k)$, and $C(h_2, k_2)$ divide $C(h, k)$ into two arcs, an upper arc and a lower arc. $P(N)$ is the union of the upper arcs so obtained. For the fractions $0/1$ and $1/1$ we use only the part of the upper arcs lying above the unit interval $[0, 1]$.

EXAMPLE. Figure 5.5 shows the path $P(3)$.

Because of Theorem 5.7, the path $P(N)$ always lies above the row of semi-circles connecting adjacent Farey fractions in F_N.

The path $P(N)$ is the contour used by Rademacher as a path of integration. It is convenient at this point to discuss the effect of a certain change of variable on each circle $C(h, k)$.

Theorem 5.8. *The transformation*

$$z = -ik^2 \left(\tau - \frac{h}{k} \right)$$

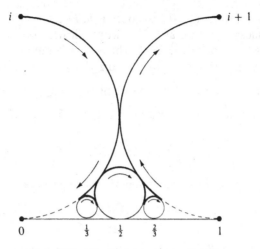

Figure 5.5   The Rademacher path $P(3)$

maps the Ford circle $C(h, k)$ in the τ-plane onto a circle K in the z-plane of radius $\frac{1}{2}$ about the point $z = \frac{1}{2}$ as center (see Figure 5.6). The points of contact $\alpha_1(h, k)$ and $\alpha_2(h, k)$ of Theorem 5.7 are mapped onto the points

$$z_1(h, k) = \frac{k^2}{k^2 + k_1{}^2} + i\frac{kk_1}{k^2 + k_1{}^2}$$

and

$$z_2(h, k) = \frac{k^2}{k^2 + k_2{}^2} - \frac{ikk_2}{k^2 + k_2{}^2}.$$

The upper arc joining $\alpha_1(h, k)$ with $\alpha_2(h, k)$ maps onto that arc of K which does not touch the imaginary z-axis.

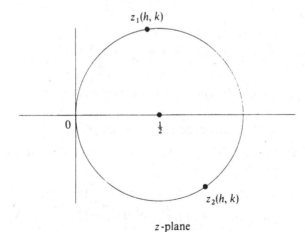

z-plane

Figure 5.6

103

PROOF. The translation $\tau - (h/k)$ moves $C(h, k)$ to the left a distance h/k, and thereby places its center at $i/(2k^2)$. Multiplication by $-ik^2$ expands the radius to $1/2$ and rotates the circle through $\pi/2$ radians in the negative direction. The expressions for $z_1(h, k)$ and $z_2(h, k)$ follow at once. $\qquad\square$

Now we obtain estimates for the moduli of z_1 and z_2.

Theorem 5.9. *For the points z_1 and z_2 of Theorem 5.8 we have*

$$(10) \qquad |z_1(h, k)| = \frac{k}{\sqrt{k^2 + k_1{}^2}}, \quad |z_2(h, k)| = \frac{k}{\sqrt{k^2 + k_2{}^2}}.$$

Moreover, if z is on the chord joining z_1 and z_2 we have

$$(11) \qquad |z| < \frac{\sqrt{2}k}{N},$$

if $h_1/k_1 < h/k < h_2/k_2$ are consecutive in F_N. The length of this chord does not exceed $2\sqrt{2}k/N$.

PROOF. For $|z_1|^2$ we have

$$|z_1|^2 = \frac{k^4 + k^2 k_1{}^2}{(k^2 + k_1{}^2)^2} = \frac{k^2}{k^2 + k_1{}^2}.$$

There is a similar formula for $|z_2|^2$. This proves (10). To prove (11) we note that if z is on the chord, then $|z| \le \max(|z_1|, |z_2|)$, so it suffices to prove that

$$(12) \qquad |z_1| < \frac{\sqrt{2}k}{N} \quad \text{and} \quad |z_2| < \frac{\sqrt{2}k}{N}.$$

For this purpose we use the inequality relating the arithmetic mean and the root mean square:

$$\frac{k + k_1}{2} \le \left(\frac{k^2 + k_1{}^2}{2}\right)^{1/2}.$$

This gives us

$$(k^2 + k_1{}^2)^{1/2} \ge \frac{k + k_1}{\sqrt{2}} \ge \frac{N + 1}{\sqrt{2}} > \frac{N}{\sqrt{2}},$$

so (10) and (12) imply (11). The length of the chord is $\le |z_1| + |z_2|$. $\qquad\square$

5.7 Rademacher's convergent series for $p(n)$

Theorem 5.10. *If $n \ge 1$ the partition function $p(n)$ is represented by the convergent series*

$$p(n) = \frac{1}{\pi\sqrt{2}} \sum_{k=1}^{\infty} A_k(n)\sqrt{k}\, \frac{d}{dn}\left(\frac{\sinh\left\{\dfrac{\pi}{k}\sqrt{\dfrac{2}{3}\left(n - \dfrac{1}{24}\right)}\right\}}{\sqrt{n - \dfrac{1}{24}}}\right)$$

where

$$A_k(n) = \sum_{\substack{0 \le h < k \\ (h, k) = 1}} e^{\pi i s(h, k) - 2\pi i n h/k}.$$

PROOF. We have

(13) $\quad p(n) = \dfrac{1}{2\pi i} \displaystyle\int_C \dfrac{F(x)}{x^{n+1}} dx \quad$ where $\quad F(x) = \displaystyle\prod_{m=1}^{\infty} (1 - x^m)^{-1} = \sum_{n=0}^{\infty} p(n)x^n;$

C is any positively oriented closed curve surrounding $x = 0$ and lying inside the unit circle. The change of variable

$$x = e^{2\pi i \tau}$$

maps the unit disk $|x| \le 1$ onto an infinite vertical strip of width 1 in the τ-plane, as shown in Figure 5.7. As x traverses counterclockwise a circle of

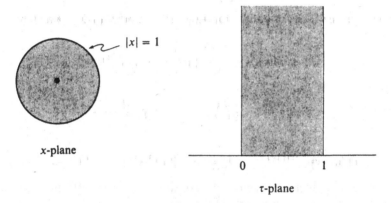

x-plane

τ-plane

Figure 5.7

radius $e^{-2\pi}$ with center at $x = 0$, the point τ varies from i to $i + 1$ along a horizontal segment. We replace this segment by the Rademacher path $P(N)$ composed of the upper arcs of the Ford circles formed for the Farey series F_N. Then (13) becomes

$$p(n) = \int_i^{i+1} F(e^{2\pi i \tau})e^{-2\pi i n \tau} \, d\tau = \int_{P(N)} F(e^{2\pi i \tau})e^{-2\pi i n \tau} \, d\tau.$$

In this discussion the integer n is kept fixed and the integer N will later be allowed to approach infinity. We can also write

$$\int_{P(N)} = \sum_{k=1}^{N} \sum_{\substack{0 \le h < k \\ (h, k) = 1}} \int_{\gamma(h, k)} = \sum_{h, k} \int_{\gamma(h, k)}$$

where $\gamma(h, k)$ denotes the upper arc of the circle $C(h, k)$, and $\sum_{h, k}$ is an abbreviation for the double sum over h and k.

Now we make the change of variable

$$z = -ik^2\left(\tau - \frac{h}{k}\right)$$

so that

$$\tau = \frac{h}{k} + \frac{iz}{k^2}.$$

Theorem 5.8 shows that this maps $C(h, k)$ onto a circle K of radius $\frac{1}{2}$ about $z = \frac{1}{2}$ as center. The arc $\gamma(h, k)$ maps onto an arc joining the points $z_1(h, k)$ and $z_2(h, k)$ in Figure 5.6. We now have

$$p(n) = \sum_{h, k} \int_{z_1(h, k)}^{z_2(h, k)} F\left(\exp\left(\frac{2\pi ih}{k} - \frac{2\pi z}{k^2}\right)\right) \frac{i}{k^2} e^{-2\pi inh/k} e^{2\pi nz/k^2} dz$$

$$= \sum_{h, k} ik^{-2} e^{-2\pi inh/k} \int_{z_1(h, k)}^{z_2(h, k)} e^{2\pi nz/k^2} F\left(\exp\left(\frac{2\pi ih}{k} - \frac{2\pi z}{k^2}\right)\right) dz.$$

Now we use the transformation formula for F (Theorem 5.1) which states that

$$F(x) = \omega(h, k)\left(\frac{z}{k}\right)^{1/2} \exp\left(\frac{\pi}{12z} - \frac{\pi z}{12k^2}\right) F(x'),$$

where

$$x = \exp\left(\frac{2\pi ih}{k} - \frac{2\pi z}{k^2}\right), \qquad x' = \exp\left(\frac{2\pi iH}{k} - \frac{2\pi}{z}\right),$$

and

$$\omega(h, k) = e^{\pi is(h, k)}, \qquad hH \equiv -1 \pmod k, \qquad (h, k) = 1.$$

Denote the elementary factor $z^{1/2} \exp[\pi/(12z) - \pi z/(12k^2)]$ by $\Psi_k(z)$ and split the integral into two parts by writing

$$F(x') = 1 + \{F(x') - 1\}.$$

We then obtain

$$p(n) = \sum_{h, k} ik^{-5/2} \omega(h, k) e^{-2\pi inh/k} (I_1(h, k) + I_2(h, k))$$

where

$$I_1(h, k) = \int_{z_1(h, k)}^{z_2(h, k)} \Psi_k(z) e^{2\pi nz/k^2} dz$$

and

$$I_2(h, k) = \int_{z_1(h, k)}^{z_2(h, k)} \Psi_k(z)\left\{F\left(\exp\left(\frac{2\pi iH}{k} - \frac{2\pi}{z}\right)\right) - 1\right\} e^{2\pi nz/k^2} dz.$$

We show next that I_2 is small for large N. The path of integration in the z-plane can be moved so that we integrate along the chord joining $z_1(h, k)$ and $z_2(h, k)$. (See Figure 5.8.) We have already estimated the length of this

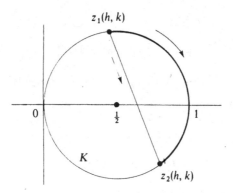

Figure 5.8

chord; it does not exceed $2\sqrt{2}k/N$. On the chord itself we have $|z|$ $\leq \max\{|z_1|, |z_2|\} < \sqrt{2}k/N$. Note also that the mapping $w = 1/z$ maps the disk bounded by K onto the half-plane $\mathrm{Re}(w) \geq 1$. Inside and on the circle K we have $0 < \mathrm{Re}(z) \leq 1$ and $\mathrm{Re}(1/z) \geq 1$, while on K itself we have $\mathrm{Re}(1/z) = 1$.

Now we estimate the integrand on the chord. We have

$$\left| \Psi_k(z) \left\{ F\left(\exp\left(\frac{2\pi i H}{k} - \frac{2\pi}{z} \right) \right) - 1 \right\} e^{2n\pi z/k^2} \right|$$

$$= |z|^{1/2} \exp\left\{ \frac{\pi}{12} \mathrm{Re}\left(\frac{1}{z}\right) - \frac{\pi}{12k^2} \mathrm{Re}(z) \right\}$$

$$\times \; e^{2n\pi \mathrm{Re}(z)/k^2} \left| \sum_{m=1}^{\infty} p(m) e^{2\pi i H m/k} e^{-2\pi m/z} \right|$$

$$\leq |z|^{1/2} \exp\left\{ \frac{\pi}{12} \mathrm{Re}\left(\frac{1}{z}\right) \right\} e^{2n\pi/k^2} \sum_{m=1}^{\infty} p(m) e^{-2\pi m \mathrm{Re}(1/z)}$$

$$< |z|^{1/2} e^{2n\pi} \sum_{m=1}^{\infty} p(m) e^{-2\pi(m - (1/24))\mathrm{Re}(1/z)}$$

$$\leq |z|^{1/2} e^{2n\pi} \sum_{m=1}^{\infty} p(m) e^{-2\pi(m - (1/24))}$$

$$= |z|^{1/2} e^{2n\pi} \sum_{m=1}^{\infty} p(m) e^{-2\pi(24m - 1)/24}$$

$$< |z|^{1/2} e^{2n\pi} \sum_{m=1}^{\infty} p(24m - 1) e^{-2\pi(24m - 1)/24}$$

$$= |z|^{1/2} e^{2n\pi} \sum_{m=1}^{\infty} p(24m - 1) y^{24m - 1} \quad \text{(where } y = e^{-2\pi/24}\text{)}$$

$$= c|z|^{1/2},$$

where

$$c = e^{2n\pi} \sum_{m=1}^{\infty} p(24m - 1)y^{24m-1}.$$

The number c does not depend on z or on N. (It depends on n, but n is fixed in this discussion.) Since z is on the chord we have $|z| < \sqrt{2}k/N$ so the integrand is bounded by $c2^{1/4}(k/N)^{1/2}$. The length of the path is less than $2\sqrt{2}k/N$, so altogether we find

$$|I_2(h, k)| < Ck^{3/2}N^{-3/2}$$

for some constant C, and therefore

$$\left|\sum_{h,k} ik^{-5/2}\omega(h, k)e^{-2\pi inh/k}I_2(h, k)\right| < \sum_{k=1}^{N} \sum_{\substack{0 \le h < k \\ (h, k) = 1}} Ck^{-1}N^{-3/2}$$

$$\le CN^{-3/2} \sum_{k=1}^{N} 1 = CN^{-1/2}.$$

This means we can write

$$(14) \qquad p(n) = \sum_{k=1}^{N} \sum_{\substack{0 \le h < k \\ (h, k) = 1}} ik^{-5/2}\omega(h, k)e^{-2\pi inh/k}I_1(h, k) + O(N^{-1/2}).$$

Next we deal with $I_1(h, k)$. This is an integral joining $z_1(h, k)$ and $z_2(h, k)$ along an arc of the circle K in Figure 5.8. We introduce the entire circle K as path of integration and show that the error made is also $O(N^{-1/2})$. We have

$$I_1(h, k) = \int_{K(-)} - \int_0^{z_1(h, k)} - \int_{z_2(h, k)}^0 = \int_{K(-)} - J_1 - J_2,$$

where $K(-)$ denotes that the integration is in the negative direction along K. To estimate $|J_1|$ we note that the length of the arc joining 0 to $z_1(h, k)$ is less than

$$\pi|z_1(h, k)| < \pi\sqrt{2}\frac{k}{N}.$$

Since $\mathrm{Re}(1/z) = 1$ and $0 < \mathrm{Re}(z) \le 1$ on K the integrand has absolute value

$$|\Psi_k(z)e^{2n\pi z/k^2}| = e^{2n\pi\mathrm{Re}(z)/k^2}|z|^{1/2} \exp\left\{\frac{\pi}{12}\mathrm{Re}\left(\frac{1}{z}\right) - \frac{\pi}{12k^2}\mathrm{Re}(z)\right\}$$

$$\le \frac{e^{2n\pi}2^{1/4}k^{1/2}e^{\pi/12}}{N^{1/2}},$$

so that

$$|J_1| < C_1k^{3/2}N^{-3/2}$$

where C_1 is a constant. A similar estimate holds for $|J_2|$ and, as before, this leads to an error term $O(N^{-1/2})$ in the formula for $p(n)$. Hence (14) becomes

$$p(n) = \sum_{\substack{k=1 \\ 0 \le h < k \\ (h,k)=1}}^{N} \sum ik^{-5/2}\omega(h,k)e^{-2\pi inh/k} \int_{K(-)} \Psi_k(z)e^{2\pi nz/k^2}\,dz + O(N^{-1/2}).$$

Now we let $N \to \infty$ to obtain

$$p(n) = i\sum_{k=1}^{\infty} A_k(n)k^{-5/2}\int_{K(-)} z^{1/2}\exp\left\{\frac{\pi}{12z} + \frac{2\pi z}{k^2}\left(n - \frac{1}{24}\right)\right\}dz,$$

where

$$A_k(n) = \sum_{\substack{0 \le h < k \\ (h,k)=1}} e^{\pi is(h,k) - 2\pi inh/k}.$$

The integral can be evaluated in terms of Bessel functions. The change of variable

$$w = \frac{1}{z}, \qquad dz = -\frac{1}{w^2}\,dw,$$

gives us

$$p(n) = \frac{1}{i}\sum_{k=1}^{\infty} A_k(n)k^{-5/2}\int_{1-\infty i}^{1+\infty i} w^{-5/2}\exp\left\{\frac{\pi w}{12} + \frac{2\pi}{k^2}\left(n - \frac{1}{24}\right)\frac{1}{w}\right\}dw.$$

Now put $t = \pi w/12$ and the formula becomes

$$p(n) = 2\pi\left(\frac{\pi}{12}\right)^{3/2}\sum_{k=1}^{\infty} A_k(n)k^{-5/2}\frac{1}{2\pi i}\int_{c-\infty i}^{c+\infty i} t^{-5/2}\exp\left\{t + \frac{\pi^2}{6k^2}\left(n - \frac{1}{24}\right)\frac{1}{t}\right\}dt$$

where $c = \pi/12$. Now on page 181 of Watson's treatise on Bessel functions [53] we find the formula

$$I_\nu(z) = \frac{(\frac{1}{2}z)^\nu}{2\pi i}\int_{c-\infty i}^{c+\infty i} t^{-\nu-1}e^{t+(z^2/4t)}\,dt \quad \text{(if } c > 0,\ \mathrm{Re}(\nu) > 0),$$

where $I_\nu(z) = i^{-\nu}J_\nu(iz)$. Taking

$$\frac{z}{2} = \left\{\frac{\pi^2}{6k^2}\left(n - \frac{1}{24}\right)\right\}^{1/2}$$

and $\nu = 3/2$ we get

$$p(n) = (2\pi)\left(\frac{\pi}{12}\right)^{3/2}\sum_{k=1}^{\infty} A_k(n)k^{-5/2}\frac{\pi^{-3/2}\left(n - \frac{1}{24}\right)^{-3/4}}{6^{-3/4}k^{-3/2}}I_{3/2}\left(\frac{\pi}{k}\sqrt{\frac{2}{3}\left(n - \frac{1}{24}\right)}\right)$$

$$= \frac{(2\pi)\left(n - \frac{1}{24}\right)^{-3/4}}{(24)^{3/4}}\sum_{k=1}^{\infty} A_k(n)k^{-1}I_{3/2}\left(\frac{\pi}{k}\sqrt{\frac{2}{3}\left(n - \frac{1}{24}\right)}\right).$$

But Bessel functions of hàlf odd order căn be reduced to elementary functions. In this case we have

$$I_{3/2}(z) = \sqrt{\frac{2z}{\pi}} \frac{d}{dz} \left(\frac{\sinh z}{z} \right).$$

Introducing this in the previous formula we finally get Rademacher's formula,

$$p(n) = \frac{1}{\pi\sqrt{2}} \sum_{k=1}^{\infty} A_k(n) k^{1/2} \frac{d}{dn} \left(\frac{\sinh\left\{ \frac{\pi}{k} \sqrt{\frac{2}{3} \left(n - \frac{1}{24} \right)} \right\}}{\sqrt{n - \frac{1}{24}}} \right). \qquad \Box$$

Exercises for Chapter 5

1. Two reduced fractions a/b and c/d are said to be *similarly ordered* if $(c - a)(d - b) \geq 0$. Let $a_1/b_1 < a_2/b_2 < \cdots$ denote the Farey fractions in F_n.
   (a) Prove that any two neighbors a_i/b_i and a_{i+1}/b_{i+1} are similarly ordered.
   (b) Prove also that any two second neighbors a_i/b_i and a_{i+2}/b_{i+2} are similarly ordered.
   Note: Erdös [8] has shown that there is an absolute constant $c > 0$ such that the kth neighbors a_i/b_i and a_{i+k}/b_{i+k} in F_n are similarly ordered if $n > ck$.

2. If a, b, c, d are positive integers such that $a/b < c/d$ and if λ and μ are positive integers, prove that the fraction

$$\theta = \frac{\lambda a + \mu c}{\lambda b + \mu d}$$

   lies between a/b and c/d, and that $(c - d\theta)/(\theta b - a) = \lambda/\mu$. When $\lambda = \mu$, θ is the mediant of a/b and c/d.

3. If $bc - ad = 1$ and $n > \max(b, d)$, prove that the terms of the Farey sequence F_n between a/b and c/d are the fractions of the form $(\lambda a + \mu c)/(\lambda b + \mu d)$ for which λ and μ are positive relatively prime integers with $\lambda b + \mu d \leq n$. Geometrically, each pair (λ, μ) is a lattice point (with coprime coordinates) in the triangle determined by the coordinate axes and the line $bx + dy = n$. Neville [29] has shown that the number of such lattice points is

$$\frac{3}{\pi^2} \frac{n^2}{bd} + O(n \log n).$$

   This shows that for a given n, the number of Farey fractions between a/b and c/d is asymptotically proportional to $1/(bd)$, the length of the interval $[a/b, c/d]$.

   Exercises 4 through 8 relate Farey fractions to lattice points in the plane. In these exercises, $n \geq 1$ and T_n denotes the set of lattice points (x, y) in the triangular region defined by the inequalities

$$1 \leq x \leq n, \qquad 1 \leq y \leq n, \qquad n + 1 \leq x + y \leq 2n.$$

Also, T_n' denotes the set of lattice points (x, y) in T_n with relatively prime coordinates.

4. Prove that a/b and c/d are consecutive fractions in the Farey sequence F_n if, and only if, the lattice point $(b, d) \in T_n'$.

5. Prove that $\sum_{(b, d) \in T_n'} 1/(bd) = 1$. *Hint*: Theorem 5.5.

6. Assign a weight $f(x, y)$ to each lattice point (x, y) and let S_n be the sum of all the weights in T_n,

$$S_n = \sum_{(x, y) \in T_n} f(x, y).$$

(a) By comparing the regions T_r and T_{r-1} for $r \geq 2$ show that

$$S_r - S_{r-1} = f(r, r) + \sum_{k=1}^{r-1} \{f(k, r) + f(r, k)\} - \sum_{k=1}^{r-1} f(k, r - k),$$

and deduce that

$$S_n = \sum_{r=1}^{n} f(r, r) + \sum_{r=2}^{n} \sum_{k=1}^{r-1} \{f(k, r) + f(r, k)\} - \sum_{r=2}^{n} \sum_{k=1}^{r-1} f(k, r - k).$$

Note: If $f(x, y) = 0$ whenever $(x, y) > 1$ this reduces to a formula of J. Lehner and M. Newman [25],

(15) $$\sum_{\substack{(x, y) \in T_n'}} f(x, y) = f(1, 1) + \sum_{r=2}^{n} \sum_{\substack{k=1 \\ (k, r) = 1}}^{r-1} \{f(k, r) + f(r, k) - f(k, r - k)\}.$$

This relates a sum involving Farey fractions to one which does not.

7. Let

$$S_n = \sum_{(b, d) \in T_n'} \frac{1}{bd(b + d)}.$$

(a) Use Exercise 5 to show that $1/(2n - 1) \leq S_n \leq 1/(n + 1)$.
(b) Choose $f(x, y) = 1/(xy(x + y))$ in (15) and show that

$$S_n = \frac{3}{2} - 2 \sum_{r=1}^{n} \sum_{\substack{k=1 \\ (k, r) = 1}}^{r} \frac{1}{r^2(r + k)}.$$

When $n \to \infty$ this gives a formula of Gupta [12],

$$\sum_{r=1}^{\infty} \sum_{\substack{k=1 \\ (k, r) = 1}}^{r} \frac{1}{r^2(r + k)} = \frac{3}{4}.$$

8. Exercise 7(a) shows that $S_n \to 0$ as $n \to \infty$. This exercise outlines a proof of the asymptotic formula

(16) $$S_n = \frac{12 \log 2}{\pi^2 n} + O\left(\frac{\log n}{n^2}\right)$$

obtained by Lehner and Newman in [25].

5: Rademacher's series for the partition function

Let

$$A_r = \sum_{\substack{k=1 \\ (k,r)=1}}^{r} \frac{1}{r^2(r+k)} = \sum_{k=1}^{r} \sum_{d \mid (r,k)} \frac{\mu(d)}{r^2(r+k)},$$

so that

$$S_n = 2 \sum_{r>n} A_r.$$

(a) Show that

$$A_r = \sum_{d \mid r} \sum_{h=1}^{d} \frac{d\mu(r/d)}{r^3(h+d)}$$

and deduce that

$$A_r = \log 2 \frac{\varphi(r)}{r^3} + O\left(\frac{1}{r^3} \sum_{d \mid r} |\mu(d)|\right).$$

(b) Show that $\sum_{r=1}^{n} \sum_{d \mid r} |\mu(d)| = O(n \log n)$ and deduce that

$$\sum_{r>n} \frac{1}{r^3} \sum_{d \mid r} |\mu(d)| = O\left(\frac{\log n}{n^2}\right).$$

(c) Use the formula $\sum_{r \le n} \varphi(r) = 3n^2/\pi^2 + O(n \log n)$ (proved in [4], Theorem 3.7) to deduce that

$$\sum_{r>n} \frac{\varphi(r)}{r^3} = \frac{6}{n\pi^2} + O\left(\frac{\log n}{n^2}\right).$$

(d) Use (a), (b), and (c) to deduce (16).

112

Modular forms with multiplicative coefficients

6

6.1 Introduction

The material in this chapter is motivated by properties shared by the discriminant $\Delta(\tau)$ and the Eisenstein series

$$G_{2k}(\tau) = \sum_{(m,n) \neq (0,0)} \frac{1}{(m + n\tau)^{2k}},$$

where k is an integer, $k \geq 2$. All these functions satisfy the relation

$$(1) \qquad f\left(\frac{a\tau + b}{c\tau + d}\right) = (c\tau + d)^r f(\tau),$$

where r is an integer and $\begin{pmatrix} a & b \\ c & d \end{pmatrix}$ is any element of the modular group Γ.
The function Δ satisfies (1) with $r = 12$, and G_{2k} satisfies (1) with $r = 2k$. Functions satisfying (1) together with some extra conditions concerning analyticity are called *modular forms*. (A precise definition is given in the next section.)

Modular forms are periodic with period 1 and have Fourier expansions. For example, we have the Fourier expansion,

$$\Delta(\tau) = (2\pi)^{12} \sum_{n=1}^{\infty} \tau(n)e^{2\pi i n\tau},$$

where $\tau(n)$ is Ramanujan's function, and

$$G_{2k}(\tau) = 2\zeta(2k) + \frac{2(2\pi i)^{2k}}{(2k - 1)!} \sum_{n=1}^{\infty} \sigma_{2k-1}(n)e^{2\pi i n\tau},$$

where $\sigma_\alpha(n)$ is the sum of the αth powers of the divisors of n.

Both $\tau(n)$ and $\sigma_\alpha(n)$ are multiplicative arithmetical functions; that is, we have

(2) $\tau(m)\tau(n) = \tau(mn)$ and $\sigma_\alpha(m)\sigma_\alpha(n) = \sigma_\alpha(mn)$ whenever $(m, n) = 1$.

They also satisfy the more general multiplicative relations

(3)
$$\tau(m)\tau(n) = \sum_{d|(m, n)} d^{11}\tau\left(\frac{mn}{d^2}\right)$$

and

(4)
$$\sigma_\alpha(m)\sigma_\alpha(n) = \sum_{d|(m, n)} d^\alpha\sigma_\alpha\left(\frac{mn}{d^2}\right)$$

for all positive integers m and n. These reduce to (2) when $(m, n) = 1$.

The striking resemblance between (3) and (4) suggests the problem of determining all modular forms whose Fourier coefficients satisfy a multiplicative property encompassing (3) and (4). The problem was solved by Hecke [16] in 1937 and his solution is discussed in this chapter.

6.2 Modular forms of weight k

In this discussion k denotes an integer (positive, negative, or zero), H denotes the upper half-plane, $H = \{\tau : \text{Im}(\tau) > 0\}$, and Γ denotes the modular group.

Definition. A function f is said to be an *entire modular form of weight k* if it satisfies the following conditions:

(a) f is analytic in the upper half-plane H.

(b) $f\left(\dfrac{a\tau + b}{c\tau + d}\right) = (c\tau + d)^k f(\tau)$ whenever $\begin{pmatrix} a & b \\ c & d \end{pmatrix} \in \Gamma$.

(c) The Fourier expansion of f has the form

$$f(\tau) = \sum_{n=0}^{\infty} c(n)e^{2\pi i n\tau}.$$

Note. The Fourier expansion of a function of period 1 is its Laurent expansion near the origin $x = 0$, where $x = e^{2\pi i \tau}$. Condition (c) states that the Laurent expansion of an entire modular form contains no negative powers of x. In other words, an entire modular form is analytic everywhere in H and at $i\infty$.

The constant term $c(0)$ is called the value of f at $i\infty$, denoted by $f(i\infty)$. If $c(0) = 0$ the function f is called a *cusp form* ("Spitzenform" in German), and the smallest r such that $c(r) \neq 0$ is called the order of the zero of f at $i\infty$. It should be noted that the discriminant Δ is a cusp form of weight 12 with a first order zero at $i\infty$. Also, no Eisenstein series G_{2k} vanishes at $i\infty$.

Warning. Some authors refer to the weight k as the "dimension $-k$" or the "degree $-k$." Others write $2k$ where we have written k.

In more general treatments a modular form is allowed to have poles in H or at $i\infty$. This is why forms satisfying our conditions are called *entire* forms. The modular function J is an example of a nonentire modular form of weight 0 since it has a pole at $i\infty$. Also, to encompass the Dedekind eta function there are extensions of the theory in which k is not restricted to integer values but may be any real number, and a factor $\varepsilon(a, b, c, d)$ of absolute value 1 is allowed in the functional equation (b). This chapter treats only entire forms of integer weight with multiplier $\varepsilon = 1$.

The zero function is a modular form of weight k for every k. A nonzero constant function is a modular form of weight k only if $k = 0$. An entire modular form of weight 0 is a modular *function* (as defined in Chapter 2) and since it is analytic everywhere in H, including the point $i\infty$, it must be constant.

Our first goal is to prove that nonconstant entire modular forms exist only if k is even and ≥ 4. Moreover, they can all be expressed in terms of the Eisenstein series G_4 and G_6. The proof is based on a formula relating the weight k with the number of zeros of f in the closure of the fundamental region of the modular group.

6.3 The weight formula for zeros of an entire modular form

We recall that the fundamental region R_Γ has vertices at the points ρ, i, $\rho + 1$ and $i\infty$. If f has a zero of order r at a point p we write $r = N(p)$.

Theorem 6.1. *Let f be an entire modular form of weight k which is not identically zero, and assume f has N zeros in the closure of the fundamental region R_Γ, omitting the vertices. Then we have the formula*

(5) $$k = 12N + 6N(i) + 4N(\rho) + 12N(i\infty).$$

PROOF. The method of proof is similar to that of Theorem 2.4 where we proved that a modular function has the same number of zeros as poles in the closure of R_Γ. Since f has no poles we can write

$$N = \frac{1}{2\pi i} \int_{\partial R} \frac{f'(\tau)}{f(\tau)} d\tau.$$

The integral is taken along the boundary of a region R formed by truncating the fundamental region by a horizontal line $y = M$ with sufficiently large M. The path ∂R is along the edges of R with circular detours made around the vertices i, ρ and $\rho + 1$ and other zeros which might occur on the edges. By calculating the limiting value of the integral as $M \to \infty$ and the circular detours shrink to their centers we find, as in the proof of Theorem 2.4,

$$(6) \qquad N = \frac{k}{12} - \frac{1}{2} N(i) - \frac{1}{3} N(\rho) - N(i\infty).$$

The only essential difference between this result and the corresponding formula obtained in the proof of Theorem 2.4 is the appearance of the term $k/12$. This comes from the weight factor $(c\tau + d)^k$ in the functional equation

$$f(A(\tau)) = (c\tau + d)^k f(\tau),$$

where $A(\tau) = (a\tau + b)/(c\tau + d)$. Differentiation of this equation gives us

$$f'(A(\tau))A'(\tau) = (c\tau + d)^k f'(\tau) + kc(c\tau + d)^{k-1} f(\tau)$$

from which we find

$$\frac{f'(A(\tau))A'(\tau)}{f(A(\tau))} = \frac{f'(\tau)}{f(\tau)} + \frac{kc}{c\tau + d}.$$

Consequently, for any path γ not passing through a zero we have

$$\frac{1}{2\pi i} \int_{A(\gamma)} \frac{f'(u)}{f(u)} du = \frac{1}{2\pi i} \int_{\gamma} \frac{f'(\tau)}{f(\tau)} d\tau + \frac{1}{2\pi i} \int_{\gamma} \frac{kc}{c\tau + d} d\tau.$$

Therefore the integrals along the arcs (2) and (3) in Figure 2.5 do not cancel as they did in the proof of Theorem 2.4 unless $k = 0$. Instead, they make a contribution whose limiting value is equal to

$$\frac{-k}{2\pi i} \int_{\rho}^{i} \frac{d\tau}{\tau} = \frac{-k}{2\pi i} (\log i - \log \rho) = \frac{-k}{2\pi i} \left(\frac{\pi i}{2} - \frac{2\pi i}{3} \right) = \frac{k}{12}.$$

The rest of the proof is like that of Theorem 2.4 and we obtain (6), which implies (5). □

From the weight formula (5) we obtain the following theorem.

Theorem 6.2

(a) *The only entire modular forms of weight $k = 0$ are the constant functions.*

(b) *If k is odd, if $k < 0$, or if $k = 2$, the only entire modular form of weight k is the zero function.*

(c) *Every nonconstant entire modular form has weight $k \geq 4$, where k is even.*

(d) *The only entire cusp form of weight $k < 12$ is the zero function.*

PROOF. Part (a) was proved earlier. To prove (b), (c) and (d) we simply refer to the weight formula in (5). Since each integer N, $N(i)$, $N(\rho)$ and $N(i\infty)$ is nonnegative, k must be nonnegative and even, with $k \geq 4$ if $k \neq 0$. Also, if $k < 12$ then $N(i\infty) = 0$ so f is not a cusp form unless $f = 0$. □

6.4 Representation of entire forms in terms of G_4 and G_6

In Chapter 1 it was shown that every Eisenstein series G_k with $k > 2$ is a polynomial in G_4 and G_6. This section shows that the same is true of every entire modular form. Since the discriminant Δ is a polynomial in G_4 and G_6,

$$\Delta = g_2{}^3 - 27g_3{}^2 = (60G_4)^3 - 27(140G_6)^2,$$

it suffices to show that all entire forms of weight k can be expressed in terms of Eisenstein series and powers of Δ. The proof repeatedly uses the fact that the product fg of two entire forms f and g of weights w_1 and w_2, respectively, is another entire form of weight $w_1 + w_2$, and the quotient f/g is an entire form of weight $w_1 - w_2$ if g has no zeros in H or at $i\infty$.

Notation. We denote by M_k the set of all entire modular forms of weight k.

Theorem 6.3. *Let f be an entire modular form of even weight $k \geq 0$ and define $G_0(\tau) = 1$ for all τ. Then f can be expressed in one and only one way as a sum of the type*

(7)
$$f = \sum_{\substack{r=0 \\ k-12r \neq 2}}^{[k/12]} a_r G_{k-12r}\Delta^r,$$

where the a_r are complex numbers. The cusp forms of even weight k are those sums with $a_0 = 0$.

PROOF. If $k < 12$ there is at most one term in the sum and the theorem can be verified directly. If f has weight $k < 12$ the weight formula (5) implies $N = N(i\infty) = 0$ so the only possible zeros of f are at the vertices ρ and i. For example, if $k = 4$ we have $N(\rho) = 1$ and $N(i) = 0$. Since G_4 has this property, f/G_4 is an entire modular form of weight 0 and therefore is constant, so $f = a_0 G_4$. Similarly, we find $f = a_0 G_k$ if $k = 6$, 8 or 10. The theorem also holds trivially for $k = 0$ (since f is constant) and for $k = 2$ (since the sum is empty). Therefore we need only consider even $k \geq 12$.

We use induction on k together with the simple observation that every cusp form in M_k can be written as a product Δh, where $h \in M_{k-12}$.

Assume the theorem has been proved for all entire forms of even weight $<k$. The form G_k has weight k and does not vanish at $i\infty$. Hence if $c = f(i\infty)/G_k(i\infty)$ the entire form $f - cG_k$ is a cusp form in M_k so $f - cG_k = \Delta h$, where $h \in M_{k-12}$. Applying the induction hypothesis to h we have

$$h = \sum_{\substack{r=0 \\ k-12-12r \neq 2}}^{[(k-12)/12]} b_r G_{k-12-12r}\Delta^r = \sum_{\substack{r=1 \\ k-12r \neq 2}}^{[k/12]} b_{r-1} G_{k-12r}\Delta^{r-1}.$$

Therefore $f = cG_k + \Delta h$ is a sum of the type shown in (7). This proves, by induction, that every entire form of even weight k has at least one representation of the type in (7). To show there is at most one such representation we need only verify that the products $G_{k-12r}\Delta^r$ are linearly independent. This follows easily from the fact that $\Delta(i\infty) = 0$ but $G_{2r}(i\infty) \neq 0$. Details are left as an exercise for the reader. $\qquad\square$

Since both Δ and G_{2r} can be expressed as polynomials in G_4 and G_6, Theorem 6.3 also shows that f is a polynomial in G_4 and G_6. The exact form of this polynomial is described in the next theorem.

Theorem 6.4. *Every entire modular form f of weight k is a polynomial in G_4 and G_6 of the type*

$$(8) \qquad\qquad f = \sum_{a,b} c_{a,b} G_4{}^a G_6{}^b$$

where the $c_{a,b}$ are complex numbers and the sum is extended over all integers $a \geq 0, b \geq 0$ such that $4a + 6b = k$.

PROOF. If k is odd, $k < 0$ or $k = 2$ the sum is empty and f is 0. If $k = 0$, f is constant and the sum consists of only one term, $c_{0,0}$. If $k = 4, 6, 8$ or 10 then each of the respective quotients f/G_4, f/G_6, $f/G_4{}^2$ and $f/(G_4 G_6)$ is an entire form of weight 0 and hence is constant. This proves (8) for $k < 12$ or k odd. To prove the result for even $k \geq 12$ we use induction on k.

Assume the theorem has been proved for all entire forms of weight $<k$. Since k is even, $k = 4m$ or $k = 4m + 2 = 4(m - 1) + 6$ for some integer $m \geq 3$. In either case there are nonnegative integers r and s such that $k = 4r + 6s$. The form $g = G_4{}^r G_6{}^s$ has weight k and does not vanish at $i\infty$. Hence if $c = f(i\infty)/g(i\infty)$ the entire form $f - cg$ is a cusp form in M_k so $f - cg = \Delta h$ where $h \in M_{k-12}$. By the induction hypothesis, h can be expressed as a sum as in (8), taken over all $a \geq 0, b \geq 0$ such that $4a + 6b = k - 12$. Multiplication by Δ gives a sum of the same type with $4a + 6b = k$. Hence $f = cg + \Delta h$ is also a sum of the required type and this proves the theorem. $\qquad\square$

6.5 The linear space M_k and the subspace $M_{k,0}$

The results of the foregoing section can be described in another way. Let M_k denote the set of all entire forms of weight k. Then M_k is a linear space over the complex field (since M_k is closed under addition and under multiplication by complex scalars). Theorem 6.3 shows that M_k is finite-dimensional with a finite basis given by the set of products $G_{k-12r}\Delta^r$ occurring in the sum (7). There are $[k/12] + 1$ terms in this sum if $k \not\equiv 2 \pmod{12}$, and one less term if $k \equiv 2 \pmod{12}$. Therefore the dimension of the space M_k is given by the formulas

(9)
$$\dim M_k = \begin{cases} \left[\dfrac{k}{12}\right] & \text{if } k \equiv 2 \pmod{12}, \\ \\ \left[\dfrac{k}{12}\right] + 1 & \text{if } k \not\equiv 2 \pmod{12}. \end{cases}$$

Another basis for M_k is the set of products $G_4{}^a G_6{}^b$ where $a \geq 0$, $b \geq 0$ and $4a + 6b = k$ (see Exercise 6.12).

The set of all cusp forms in M_k is a linear subspace of M_k which we denote by $M_{k,0}$. The representation in Theorem 6.3 shows that

(10)
$$\dim M_{k,0} = \dim M_k - 1$$

since the cusp forms are those sums in (7) with $a_0 = 0$.

We also note that if $k \geq 12$, $f \in M_{k,0}$ if and only if $f = \Delta h$, where $h \in M_{k-12}$. Therefore the linear transformation $T : M_{k-12} \to M_{k,0}$ defined by

$$T(h) = \Delta h$$

establishes an isomorphism between $M_{k,0}$ and M_{k-12}. Consequently, if $k \geq 12$ we have

(11)
$$\dim M_{k,0} = \dim M_{k-12}.$$

The two formulas (11) and (10) imply

$$\dim M_k = 1 + \dim M_{k-12}$$

if $k \geq 12$. This equation, together with the fact that $\dim M_k = 1, 0, 1, 1, 1, 1$ when $k = 0, 2, 4, 6, 8, 10$, gives another proof of (9).

EXAMPLES. Formula (9) shows that

$$\dim M_k = 1 \quad \text{if } k = 4, 6, 8, 10, \text{ and } 14.$$

Corresponding basis elements are G_4, G_6, $G_4{}^2$, $G_4 G_6$, and $G_4{}^2 G_6$.
Formulas (11) and (9) together show that

$$\dim M_{k,0} = 1 \quad \text{if } k = 12, 16, 18, 20, 22, \text{ and } 26.$$

Corresponding basis elements are Δ, ΔG_4, ΔG_6, $\Delta G_4{}^2$, $\Delta G_4 G_6$, and $\Delta G_4{}^2 G_6$.

6.6 Classification of entire forms in terms of their zeros

The next theorem gives another way of expressing all entire forms in terms of G_4, G_6, Δ and Klein's modular invariant J.

Theorem 6.5. *Let f be an entire form of weight k and let $z_1, \ldots, z_N$ denote the N zeros of f in the closure of R_Γ (omitting the vertices) with zeros of order*

119

$N(\rho)$, $N(i)$ and $N(i\infty)$ at the vertices. Then there is a constant c such that

(12) $\qquad f(\tau) = cG_4(\tau)^{N(\rho)}G_6(\tau)^{N(i)}\Delta(\tau)^{N(i\infty)}\Delta(\tau)^N \prod_{k=1}^{N} \{J(\tau) - J(z_k)\}.$

PROOF. The product

$$g(\tau) = \prod_{k=1}^{N} \{J(\tau) - J(z_k)\}$$

is a modular function with its only zeros in the closure of R_Γ at $z_1, \ldots, z_N$ and with a pole of order N at $i\infty$. Since Δ has a first-order zero at $i\infty$, the product $\Delta^N g$ is an entire modular form of weight $12N$ which, in the closure of R_Γ, vanishes only at $z_1, \ldots, z_N$. Therefore the product

$$h = G_4{}^{N(\rho)}G_6{}^{N(i)}\Delta^{N(i\infty)}\Delta^N g$$

has exactly the same zeros as f in the closure of R_Γ. Also, h is an entire modular form having the same weight as f since

$$k = 4N(\rho) + 6N(i) + 12N(i\infty) + 12N.$$

Therefore f/h is an entire form of weight 0 so f/h is constant. This proves (12). $\qquad\square$

6.7 The Hecke operators T_n

Hecke determined all entire forms with multiplicative coefficients by introducing a sequence of linear operators T_n, $n = 1, 2, \ldots$, which map the linear space M_k onto itself. Hecke's operators are defined as follows.

Definition. For a fixed integer k and any $n = 1, 2, \ldots$, the operator T_n is defined on M_k by the equation

(13) $\qquad (T_n f)(\tau) = n^{k-1} \sum_{d|n} d^{-k} \sum_{b=0}^{d-1} f\left(\frac{n\tau + bd}{d^2}\right).$

In the special case when n is prime, say $n = p$, the sum on d contains only two terms and the definition reduces to the formula

(14) $\qquad (T_p f)(\tau) = p^{k-1} f(p\tau) + \frac{1}{p} \sum_{b=0}^{p-1} f\left(\frac{\tau + b}{p}\right).$

The sum on b is the operator encountered in Chapter 4. It maps functions automorphic under Γ onto functions automorphic under the congruence subgroup $\Gamma_0(p)$.

We will show that T_n maps each f in M_k onto another function in M_k. First we describe the action of T_n on the Fourier expansion of f.

Theorem 6.6. *If $f \in M_k$ and has the Fourier expansion*

$$f(\tau) = \sum_{m=0}^{\infty} c(m)e^{2\pi im\tau},$$

then $T_n f$ has the Fourier expansion

(15)
$$(T_n f)(\tau) = \sum_{m=0}^{\infty} \gamma_n(m)e^{2\pi im\tau},$$

where

(16)
$$\gamma_n(m) = \sum_{d|(n, m)} d^{k-1} c\left(\frac{mn}{d^2}\right).$$

PROOF. From the definition in (13) we find

$$(T_n f)(\tau) = n^{k-1} \sum_{d|n} d^{-k} \sum_{b=0}^{d-1} \sum_{m=0}^{\infty} c(m)e^{2\pi im(n\tau + bd)/d^2}$$

$$= \sum_{m=0}^{\infty} \sum_{d|n} \left(\frac{n}{d}\right)^{k-1} c(m)e^{2\pi imn\tau/d^2} \frac{1}{d} \sum_{b=0}^{d-1} e^{2\pi imb/d}.$$

The sum on b is a geometric sum which is equal to d if $d | m$, and is 0 otherwise. Hence

$$(T_n f)(\tau) = \sum_{m=0}^{\infty} \sum_{d|n, d|m} \left(\frac{n}{d}\right)^{k-1} c(m)e^{2\pi imn\tau/d^2}.$$

Writing $m = qd$ we have

$$(T_n f)(\tau) = \sum_{q=0}^{\infty} \sum_{d|n} \left(\frac{n}{d}\right)^{k-1} c(qd)e^{2\pi iqn\tau/d}.$$

In the sum on d we can replace d by n/d to obtain

$$(T_n f)(\tau) = \sum_{q=0}^{\infty} \sum_{d|n} d^{k-1} c\left(\frac{qn}{d}\right) e^{2\pi iqd\tau}.$$

If $x = e^{2\pi i\tau}$ the last sum contains powers of the form x^{qd}. We collect those terms for which qd is constant, say $qd = m$. Then $q = m/d$ and $d | m$ so

$$(T_n f)(\tau) = \sum_{m=0}^{\infty} \sum_{d|n, d|m} d^{k-1} c\left(\frac{mn}{d^2}\right) x^m,$$

which implies (16). $\qquad\square$

Our next task is to prove that T_n maps M_k into itself. For this purpose we note that the definition of $T_n f$ can be written in a slightly different form. We write $n = ad$ and let

$$A\tau = \frac{a\tau + b}{d}.$$

Then (13) takes the form

$$(17) \qquad (T_n f)(\tau) = n^{k-1} \sum_{\substack{a \geq 1, \, ad = n \\ 0 \leq b < d}} d^{-k} f(A\tau) = \frac{1}{n} \sum_{\substack{a \geq 1, \, ad = n \\ 0 \leq b < d}} a^k f(A\tau).$$

The matrix $\begin{pmatrix} a & b \\ 0 & d \end{pmatrix}$ which represents A has determinant $ad = n$. To determine the behavior of $T_n f$ under transformations of the modular group Γ we need some properties of transformations with determinant n. These are described in the next section.

6.8 Transformations of order n

Let n be a fixed positive integer. A transformation of the form

$$A\tau = \frac{a\tau + b}{c\tau + d},$$

where a, b, c, d are integers with $ad - bc = n$, is called a *transformation of order n*. It can be represented by the 2×2 matrix

$$A = \begin{pmatrix} a & b \\ c & d \end{pmatrix}$$

where, as usual, we identify each matrix with its negative.

We denote by $\Gamma(n)$ the set of all transformations of order n. The modular group Γ is $\Gamma(1)$.

Two transformations A_1 and A_2 in $\Gamma(n)$ are called *equivalent*, and we write $A_1 \sim A_2$, if there is a transformation V in Γ such that

$$A_1 = V A_2.$$

The relation $\sim$ is obviously reflexive, symmetric, and transitive, and hence is an equivalence relation. Consequently, the set $\Gamma(n)$ can be partitioned into equivalence classes such that two elements of $\Gamma(n)$ are in the same class if, and only if, they are equivalent. The next theorem describes a set of representatives.

Theorem 6.7. *In every equivalence class of $\Gamma(n)$ there is a representative of triangular form*

$$\begin{pmatrix} a & b \\ 0 & d \end{pmatrix}, \quad \text{where } d > 0.$$

PROOF. Let $A = \begin{pmatrix} a & b \\ c & d \end{pmatrix}$ be an arbitrary element of $\Gamma(n)$. If $c = 0$ there is nothing more to prove. If $c \neq 0$ we reduce the fraction $-a/c$ to lowest terms. That is, we choose integers r and s such that $s/r = -a/c$ and $(r, s) = 1$.

Next we choose two integers p and q such that $ps - qr = 1$ and let

$$V = \begin{pmatrix} p & q \\ r & s \end{pmatrix}.$$

Then $V \in \Gamma$ and

$$VA = \begin{pmatrix} p & q \\ r & s \end{pmatrix} \begin{pmatrix} a & b \\ c & d \end{pmatrix} = \begin{pmatrix} pa + qc & pb + qd \\ ra + sc & rb + sd \end{pmatrix}.$$

Since $ra + sc = 0$ and $\det(VA) = \det V \det A = n$ we see that $VA \in \Gamma(n)$ so $VA \sim A$. Hence VA or its negative is the required representative. $\qquad\square$

Theorem 6.8. *A complete system of nonequivalent elements in $\Gamma(n)$ is given by the set of transformations of triangular form*

(18)
$$A = \begin{pmatrix} a & b \\ 0 & d \end{pmatrix},$$

where d runs through the positive divisors of n and, for each fixed d, $a = n/d$, and b runs through a complete residue system modulo d.

PROOF. Theorem 6.7 shows that every element in $\Gamma(n)$ is equivalent to one of the transformations in (18). Therefore we need only show that two such transformations, say

$$A_1 = \begin{pmatrix} a_1 & b_1 \\ 0 & d_1 \end{pmatrix} \quad \text{and} \quad A_2 = \begin{pmatrix} a_2 & b_2 \\ 0 & d_2 \end{pmatrix}$$

are equivalent if, and only if,

(19) $\qquad a_1 = a_2, d_1 = d_2, \quad$ and $\quad b_1 \equiv b_2 \pmod{d_1}$.

If (19) holds then $b_2 = b_1 + qd_1$ for some integer q and we can take

$$V = \begin{pmatrix} 1 & q \\ 0 & 1 \end{pmatrix}.$$

Then $VA_1 = A_2$ so $A_1 \sim A_2$.

Conversely, if $A_1 \sim A_2$ there is an element

$$V = \begin{pmatrix} p & q \\ r & s \end{pmatrix}$$

in Γ such that $A_2 = VA_1$. Therefore

(20)
$$\begin{pmatrix} a_2 & b_2 \\ 0 & d_2 \end{pmatrix} = \begin{pmatrix} p & q \\ r & s \end{pmatrix} \begin{pmatrix} a_1 & b_1 \\ 0 & d_1 \end{pmatrix} = \begin{pmatrix} pa_1 & pb_1 + qd_1 \\ ra_1 & rb_1 + sd_1 \end{pmatrix}.$$

Equating entries we find $ra_1 = 0$ so $r = 0$ since $a_1 \neq 0$ because $a_1 d_1 = n \geq 1$. Now $ps - qr = 1$ so $ps = 1$ hence both p and s are 1 or both are -1. We can assume $p = s = 1$ (otherwise replace V by $-V$). Equating the remaining entries in (20) we find $a_2 = a_1, d_2 = d_1, b_2 = b_1 + qd_1$, so $b_2 \equiv b_1 \pmod{d_1}$. This completes the proof. $\qquad\square$

Note. The sum in (17) defining $T_n f$ can now be written in the form

(21) $$(T_n f)(\tau) = \frac{1}{n} \sum_A a^k f(A\tau),$$

where A runs through a complete set of nonequivalent elements in $\Gamma(n)$ of the form described in Theorem 6.8. The coefficient a^k is the kth power of the entry in the first row and first column of A.

Theorem 6.9. *If $A_1 \in \Gamma(n)$ and $V_1 \in \Gamma$, then there exist matrices A_2 in $\Gamma(n)$ and V_2 in Γ such that*

(22) $$A_1 V_1 = V_2 A_2.$$

Moreover, if

$$A_i = \begin{pmatrix} a_i & b_i \\ 0 & d_i \end{pmatrix} \quad \text{and} \quad V_i = \begin{pmatrix} \alpha_i & \beta_i \\ \gamma_i & \delta_i \end{pmatrix}$$

for $i = 1, 2$, then we have

(23) $$a_1(\gamma_2 A_2 \tau + \delta_2) = a_2(\gamma_1 \tau + \delta_1).$$

PROOF. Since $\det(A_1 V_1) = \det A_1 \det V_1 = n$, the matrix $A_1 V_1$ is in $\Gamma(n)$ so, by Theorem 6.7, there exists A_2 in $\Gamma(n)$ and V_2 in Γ such that (22) holds. To verify (23) we first note that $A_1 V_1$ has the form

$$A_1 V_1 = \begin{pmatrix} a_1 & b_1 \\ 0 & d_1 \end{pmatrix} \begin{pmatrix} \alpha_1 & \beta_1 \\ \gamma_1 & \delta_1 \end{pmatrix} = \begin{pmatrix} * & * \\ d_1 \gamma_1 & d_1 \delta_1 \end{pmatrix}$$

and that

$$A_2^{-1} = \frac{1}{n} \begin{pmatrix} d_2 & -b_2 \\ 0 & a_2 \end{pmatrix}.$$

Therefore (22) implies

$$V_2 = A_1 V_1 A_2^{-1} = \frac{1}{n} \begin{pmatrix} * & * \\ d_1 \gamma_1 & d_1 \delta_1 \end{pmatrix} \begin{pmatrix} d_2 & -b_2 \\ 0 & a_2 \end{pmatrix}$$

$$= \frac{1}{n} \begin{pmatrix} * & * \\ d_1 d_2 \gamma_1 & -d_1 \gamma_1 b_2 + d_1 \delta_1 a_2 \end{pmatrix}.$$

Equating entries in the second row we find

$$\gamma_2 = \frac{d_1 d_2 \gamma_1}{n} = \frac{d_2}{a_1} \gamma_1$$

and

$$\delta_2 = \frac{-d_1 \gamma_1 b_2 + d_1 \delta_1 a_2}{n} = -\frac{b_2}{a_1} \gamma_1 + \frac{a_2}{a_1} \delta_1$$

since $a_1 d_1 = n$. Hence

$$a_1 \gamma_2 = d_2 \gamma_1 \quad \text{and} \quad a_1 \delta_2 = -b_2 \gamma_1 + a_2 \delta_1,$$

and we obtain

$$a_1(\gamma_2 A_2 \tau + \delta_2) = a_1 \gamma_2 A_2 \tau + a_1 \delta_2$$

$$= d_2 \gamma_1 \frac{a_2 \tau + b_2}{d_2} - b_2 \gamma_1 + a_2 \delta_1 = a_2(\gamma_1 \tau + \delta_1),$$

which proves (23). $\qquad\square$

6.9 Behavior of $T_n f$ under the modular group

Theorem 6.10. *If $f \in M_k$ and $V = \begin{pmatrix} \alpha & \beta \\ \gamma & \delta \end{pmatrix} \in \Gamma$ then*

$$(24) \qquad (T_n f)(V\tau) = (\gamma\tau + \delta)^k (T_n f)(\tau).$$

PROOF. We use the representation in (21) to write

$$(T_n f)(\tau) = \frac{1}{n} \sum_{A_1} a_1{}^k f(A_1 \tau)$$

where $A_1 = \begin{pmatrix} a_1 & b_1 \\ 0 & d_1 \end{pmatrix}$ and A_1 runs through a complete set of nonequivalent elements in $\Gamma(n)$. Replacing τ by $V\tau$ we find

$$(25) \qquad (T_n f)(V\tau) = \frac{1}{n} \sum_{A1} a_1{}^k f(A_1 V\tau).$$

By Theorems 6.7 and 6.9, there exist matrices

$$A_2 = \begin{pmatrix} a_2 & b_2 \\ 0 & d_2 \end{pmatrix} \text{in } \Gamma(n) \quad \text{and} \quad V_2 = \begin{pmatrix} \alpha_2 & \beta_2 \\ \gamma_2 & \delta_2 \end{pmatrix} \text{in } \Gamma$$

such that

$$A_1 V = V_2 A_2 \quad \text{and} \quad a_1(\gamma_2 A_2 \tau + \delta_2) = a_2(\gamma\tau + \delta).$$

Therefore

$$a_1{}^k f(A_1 V\tau) = a_1{}^k f(V_2 A_2 \tau) = a_1{}^k (\gamma_2 A_2 \tau + \delta_2)^k f(A_2 \tau)$$

$$= a_2{}^k (\gamma\tau + \delta)^k f(A_2 \tau)$$

since $f \in M_k$. Now as A_1 runs through a complete set of nonequivalent elements of $\Gamma(n)$ so does A_2. Hence (25) becomes

$$(T_n f)(V\tau) = \frac{1}{n}(\gamma\tau + \delta)^k \sum_{A_2} a_2{}^k f(A_2 \tau) = (\gamma\tau + \delta)^k (T_n f)(\tau). \qquad\square$$

The next theorem shows that each Hecke operator T_n maps M_k into M_k and also maps $M_{k,0}$ into $M_{k,0}$.

Theorem 6.11. *If $f \in M_k$ then $T_n f \in M_k$. Moreover, if f is a cusp form then $T_n f$ is also a cusp form.*

PROOF. If $f \in M_k$ the definition of T_n shows that $T_n f$ is analytic everywhere in H. Theorem 6.6 shows that $T_n f$ has a Fourier expansion of the required form and that $T_n f$ is analytic at $i\infty$. And Theorem 6.10 shows that $T_n f$ has the proper behavior under transformations of Γ. Finally, if f is a cusp form, the Fourier expansion in Theorem 6.6 shows that $T_n f$ is also a cusp form. □

6.10 Multiplicative property of Hecke operators

This section shows that any two Hecke operators T_n and T_m defined on M_k commute with each other. This follows from a multiplicative property of the composition $T_m T_n$. First we treat the case in which m and n are relatively prime.

Theorem 6.12. *If $(m, n) = 1$ we have the composition property*

$$(26) \qquad T_m T_n = T_{mn}.$$

PROOF. If $f \in M_k$ we have

$$(T_n f)(\tau) = \frac{1}{n} \sum_{\substack{a \geq 1, \, ad = n \\ 0 \leq b < d}} a^k f(A\tau),$$

where $A = \begin{pmatrix} a & b \\ 0 & d \end{pmatrix}$. Applying T_m to each member we have

$$\{T_m(T_n(f))\}(\tau) = \frac{1}{m} \sum_{\substack{\alpha \geq 1, \, \alpha\delta = m \\ 0 \leq \beta < \delta}} \alpha^k \frac{1}{n} \sum_{\substack{a \geq 1, \, ad = n \\ 0 \leq b < d}} a^k f(BA\tau),$$

where $B = \begin{pmatrix} \alpha & \beta \\ 0 & \delta \end{pmatrix}$. This can be written as

$$(27) \qquad \{(T_m T_n)(f)\}(\tau) = \frac{1}{mn} \sum_{\substack{\alpha \geq 1, \, \alpha\delta = m \\ 0 \leq \beta < \delta}} \sum_{\substack{a \geq 1, \, ad = n \\ 0 \leq b < d}} (\alpha a)^k f(C\tau),$$

where

$$C = BA = \begin{pmatrix} \alpha & \beta \\ 0 & \delta \end{pmatrix}\begin{pmatrix} a & b \\ 0 & d \end{pmatrix} = \begin{pmatrix} \alpha a & \alpha b + \beta d \\ 0 & d\delta \end{pmatrix}.$$

As d and δ run through the positive divisors of n and m, respectively, the product $d\delta$ runs through the positive divisors of mn since $(m, n) = 1$. The linear combination $\alpha b + \beta d$ runs through a complete residue system mod $d\delta$ as b and β run through complete residue systems mod d and δ, respectively. Therefore the matrix C runs through a complete set of non-equivalent elements of $\Gamma(mn)$ and we see that (27) implies (26). □

The next theorem extends the composition property in (26) to arbitrary m and n. For convenience in notation we write $T(n)$ in place of T_n.

Theorem 6.13. *Any two Hecke operators $T(n)$ and $T(m)$ defined on M_k commute with each other. Moreover, we have the composition formula*

$$(28) \qquad T(m)T(n) = \sum_{d|(m, n)} d^{k-1} T(mn/d^2).$$

PROOF. Commutativity follows from (28) since the right member is symmetric in m and n. If $(m, n) = 1$ formula (28) reduces to (26). Therefore, to prove (28) it suffices to treat the case when m and n are powers of the same prime p. First we consider the case $m = p$ and $n = p^r$, where $r \geq 1$. In this case we are to prove that

$$(29) \qquad T(p)T(p^r) = T(p^{r+1}) + p^{k-1} T(p^{r-1}).$$

We use the representation in (17) and note that the divisors of p^r have the form p^t where $0 \leq t \leq r$. Hence we have

$$(30) \qquad \{T(p^r)f\}(\tau) = p^{-r} \sum_{\substack{0 \leq t \leq r \\ 0 \leq b_t < p^t}} p^{(r-t)k} f\left(\frac{p^{r-t}\tau + b_t}{p^t}\right).$$

By (14) we have

$$\{T(p)g\}(\tau) = p^{k-1}g(p\tau) + p^{-1} \sum_{b=0}^{p-1} g\left(\frac{\tau + b}{p}\right),$$

so when we apply $T(p)$ to each member of (30) we find

$$\{T(p)T(p^r)f\}(\tau) = p^{k-1-r} \sum_{\substack{0 \leq t \leq r \\ 0 \leq b_t < p^t}} p^{(r-t)k} f\left(\frac{p^{r+1-t}\tau + pb_t}{p^t}\right)$$

$$+ p^{-1-r} \sum_{\substack{0 \leq t \leq r \\ 0 \leq b_t < p^t}} p^{(r-t)k} \sum_{b=0}^{p-1} f\left(\frac{p^{r-t}\tau + b_t + bp^t}{p^{t+1}}\right).$$

In the second sum the linear combination $b_t + bp^t$ runs through a complete residue system mod p^{t+1}. Since $r - t = (r + 1) - (t + 1)$ the second sum, together with the term $t = 0$ from the first sum, is equal to $\{T(p^{r+1})f\}(\tau)$. In the remaining terms we cancel a factor p in the argument of f, then transfer the factor p^k to each summand to obtain

$$\{T(p)T(p^r)f\}(\tau) = \{T(p^{r+1})f\}(\tau) + p^{-1-r} \sum_{\substack{1 \leq t \leq r \\ 0 \leq b_t < p^t}} p^{(r+1-t)k} f\left(\frac{p^{r-t}\tau + b_t}{p^{t-1}}\right).$$

Dividing each b_t by p^{t-1} we can write

$$b_t = q_t p^{t-1} + r_t,$$

where $0 \leq r_t < p^{t-1}$ and q_t runs through a complete residue system mod p. Since f is periodic with period 1 we have

$$f\left(\frac{p^{r-t}\tau + b_t}{p^{t-1}}\right) = f\left(\frac{p^{r-t}\tau + r_t}{p^{t-1}}\right),$$

so as q_t runs through a complete residue system mod p each term is repeated p times. Replacing the index t by $t - 1$ we see that the last sum is p^{k-1} times the sum defining $\{T(p^{r-1})f\}(\tau)$. This proves (29).

Now we consider general powers of the same prime, say $m = p^s$ and $n = p^r$. Without loss of generality we can assume that $r \leq s$. We will use induction on r to prove that

$$(31) \qquad T(p^r)T(p^s) = \sum_{t=0}^{r} p^{t(k-1)}T(p^{r+s-2t}) = \sum_{d|(p^r, p^s)} d^{k-1}T\left(\frac{p^{r+s}}{d^2}\right)$$

for all r and all $s \geq r$. When $r = 1$, (31) follows for all $s \geq 1$ from (29). Therefore we assume that (31) holds for r and all smaller powers and all $s \geq r$, and prove it also holds for $r + 1$ and all $s \geq r + 1$.

By (29) we have

$$T(p)T(p^r)T(p^s) = T(p^{r+1})T(p^s) + p^{k-1}T(p^{r-1})T(p^s),$$

and by the induction hypothesis we have

$$T(p)T(p^r)T(p^s) = \sum_{t=0}^{r} p^{t(k-1)}T(p)T(p^{r+s-2t}).$$

Equating the two expressions, solving for $T(p^{r+1})T(p^s)$ and using (29) in the sum on t we find

$$T(p^{r+1})T(p^s) = \sum_{t=0}^{r} p^{t(k-1)}T(p^{r+s+1-2t}) + \sum_{t=0}^{r} p^{(t+1)(k-1)}T(p^{r+s-1-2t})$$

$$- p^{k-1}T(p^{r-1})T(p^s).$$

By the induction hypothesis the last term cancels the second sum over t except for the term with $t = r$. Therefore

$$T(p^{r+1})T(p^s) = \sum_{t=0}^{r} p^{t(k-1)}T(p^{r+s+1-2t}) + p^{(r+1)(k-1)}T(p^{s-1-r})$$

$$= \sum_{t=0}^{r+1} p^{t(k-1)}T(p^{r+1+s-2t}).$$

This proves (31) by induction for all r and all $s \geq r$, and also completes the proof of (28). $\qquad \square$

6.11 Eigenfunctions of Hecke operators

In Theorem 6.6 we proved that if $f \in M_k$ and has the Fourier expansion

(32)
$$f(\tau) = \sum_{m=0}^{\infty} c(m)x^m,$$

where $x = e^{2\pi i \tau}$, then $T_n f$ has the Fourier expansion

(33)
$$(T_n f)(\tau) = \sum_{m=0}^{\infty} \gamma_n(m)x^m,$$

where

(34)
$$\gamma_n(m) = \sum_{d|(n, m)} d^{k-1} c\left(\frac{mn}{d^2}\right).$$

When $m = 0$ we have $(n, 0) = n$ so the constant terms of f and $T_n f$ are related by the equation

(35)
$$\gamma_n(0) = \sum_{d|n} d^{k-1} c(0) = \sigma_{k-1}(n)c(0)$$

for all $n \geq 1$. Similarly, when $m = 1$ we find

(36)
$$\gamma_n(1) = c(n)$$

for all $n \geq 1$.

The sum on the right of (34) resembles that which occurs in the multiplicative property of Ramanujan's function $\tau(n)$ and the divisor functions $\sigma_\alpha(n)$. These examples suggest we seek those forms f for which the transformed function $T_n f$ has Fourier coefficients

(37)
$$\gamma_n(m) = c(n)c(m)$$

since this would imply the multiplicative property

$$c(n)c(m) = \sum_{d|(n, m)} d^{k-1} c\left(\frac{mn}{d^2}\right).$$

The relation (37) is equivalent to the identity

$$T_n f = c(n) f$$

for all $n \geq 1$. A nonzero function f satisfying a relation of the form

(38)
$$T_n f = \lambda(n) f$$

for some complex scalar $\lambda(n)$ is called an *eigenfunction* (or *eigenform*) of the operator T_n, and the scalar $\lambda(n)$ is called an *eigenvalue* of T_n. If f is an eigenform so is cf for every $c \neq 0$.

129

EXAMPLES. If a linear operator T maps a 1-dimensional function space V into itself, then every nonzero function in V is an eigenfunction of T. Formula (9) shows that

$$\dim M_k = 1 \quad \text{if } k = 4, 6, 8, 10 \text{ and } 14,$$

so each Hecke operator T_n has eigenforms in M_k for each of these values of k. For example, the respective Eisenstein series G_4, G_6, G_8, G_{10} and G_{14} are eigenforms for each T_n.

Similarly, formula (11) implies that

$$\dim M_{k,0} = 1 \quad \text{if } k = 12, 16, 18, 20, 22 \text{ and } 26,$$

so each T_n has eigenforms in $M_{k,0}$ for each of these values of k. The respective cusp forms Δ, ΔG_4, ΔG_6, ΔG_8, ΔG_{10} and ΔG_{14} are eigenforms for each T_n.

If f is an eigenform for every Hecke operator T_n, $n \geq 1$, then f is called a *simultaneous eigenform*. All the examples just mentioned are simultaneous eigenforms.

6.12 Properties of simultaneous eigenforms

Theorem 6.14. *Assume k is even, $k \geq 4$. If the space M_k contains a simultaneous eigenform f with Fourier expansion (32), then $c(1) \neq 0$.*

PROOF. The coefficient of x in the Fourier expansion of $T_n f$ is $\gamma_n(1) = c(n)$. Since f is a simultaneous eigenform this coefficient is also equal to $\lambda(n)c(1)$, so

$$c(n) = \lambda(n)c(1)$$

for all $n \geq 1$. If $c(1) = 0$ then $c(n) = 0$ for all $n \geq 1$ and $f(\tau) = c(0)$. But then $c(0) = 0$ since $k \geq 4$, hence $f = 0$, contradicting the definition of eigenform. This proves that $c(1) \neq 0$. $\qquad\square$

An eigenform with $c(1) = 1$ is said to *normalized*. If M_k contains a simultaneous eigenform then it also contains a normalized eigenform since we can always make $c(1) = 1$ by multiplying f by a suitable nonzero constant.

It is easy to characterize all cusp forms which are simultaneous eigenforms. Since the zero function is the only cusp form of weight < 12 we need consider only $k \geq 12$.

Theorem 6.15. *Assume $f \in M_{k,0}$ where k is even, $k \geq 12$. Then f is a simultaneous normalized eigenform if, and only if, the coefficients in the Fourier expansion (32) satisfy the multiplicative property*

$$(39) \qquad c(m)c(n) = \sum_{d \mid (n, m)} d^{k-1} c\left(\frac{mn}{d^2}\right)$$

for all $m \geq 1$, $n \geq 1$, in which case the coefficient $c(n)$ is an eigenvalue of T_n.

PROOF. The equation $T_n f = \lambda(n) f$ is equivalent to the relation

(40) $$\gamma_n(m) = \lambda(n) c(m)$$

obtained by equating coefficients of x^m in the corresponding Fourier expansions. Since f is a cusp form so is $T_n f$ hence (40) is to hold for all $m \geq 1$ and $n \geq 1$. Now $\gamma_n(1) = c(n)$ so (40) implies $\lambda(n) = c(n)$ if $c(1) = 1$, and hence $\gamma_n(m) = c(n) c(m)$. On the other hand, Equation (34) shows that (40) is equivalent to (39) if $c(1) = 1$. Therefore f is a normalized simultaneous eigenform if, and only if, (39) holds for all $m \geq 1, n \geq 1$. $\qquad\square$

6.13 Examples of normalized simultaneous eigenforms

The discriminant Δ is a cusp form with Fourier expansion

$$\Delta(\tau) = (2\pi)^{12} \sum_{m=1}^{\infty} \tau(m) x^m$$

where $\tau(1) = 1$. Therefore $(2\pi)^{-12} \Delta(\tau)$ is a normalized eigenform for each T_n with corresponding eigenvalue $\tau(n)$. This also proves that Ramanujan's function $\tau(n)$ satisfies the multiplicative property in (3).

The next theorem shows that the only simultaneous eigenforms in M_{2k} which are not cusp forms are constant multiples of the Eisenstein series G_{2k}.

Theorem 6.16. *Assume that $f \in M_{2k}$, where $k \geq 2$, and that f is not a cusp form. Then f is a normalized simultaneous eigenform if, and only if,*

(41) $$f(\tau) = \frac{(2k-1)!}{2(2\pi i)^{2k}} G_{2k}(\tau).$$

PROOF. In the Fourier expansion (32) we have $c(0) \neq 0$ since f is not a cusp form. The relation

(42) $$T_n f = \lambda(n) f$$

is equivalent to the relation

(43) $$\gamma_n(m) = \lambda(n) c(m)$$

obtained by equating coefficients of x^m in the corresponding Fourier expansions. When $m = 0$ this becomes

$$\gamma_n(0) = \lambda(n) c(0).$$

On the other hand, (35) implies $\gamma_n(0) = \sigma_{2k-1}(n) c(0)$ since $f \in M_{2k}$. But $c(0) \neq 0$, so Equation (42) holds if, and only if,

$$\lambda(n) = \sigma_{2k-1}(n).$$

Using this in (43) we find that

$$\gamma_n(m) = \sigma_{2k-1}(n)c(m).$$

When $m = 1$ this relation, together with (36), gives us

$$c(n) = \sigma_{2k-1}(n)c(1).$$

Therefore, f is a normalized simultaneous eigenform in M_{2k} if, and only if,

$$c(n) = \sigma_{2k-1}(n)$$

for all $n \geq 1$. Since the Eisenstein series G_{2k} has the Fourier expansion

$$G_{2k}(\tau) = 2\zeta(2k) + \frac{2(2\pi i)^{2k}}{(2k-1)!} \sum_{m=1}^{\infty} \sigma_{2k-1}(m)x^m,$$

the function in (41) is normalized and its Fourier expansion is given by

(44) $$f(\tau) = \frac{(2k-1)!}{(2\pi i)^{2k}} \zeta(2k) + \sum_{m=1}^{\infty} \sigma_{2k-1}(m)x^m. \qquad \square$$

Note. Since

$$\zeta(2k) = (-1)^{k+1} \frac{(2\pi)^{2k}}{2(2k)!} B_{2k}$$

where B_k is the kth Bernoulli number defined by

$$\frac{x}{e^x - 1} = \sum_{k=0}^{\infty} \frac{B_k}{k!} x^k,$$

the constant term in (44) is equal to $-B_{2k}/(4k)$. (See [4], Theorem 12.17.)
    We can also write

$$G_{2k}(\tau) = 2\zeta(2k)\left\{ 1 - \frac{4k}{B_{2k}} \sum_{m=1}^{\infty} \sigma_{2k-1}(m)x^m \right\}.$$

Since the eigenvalue $\lambda(n)$ in (42) is $\sigma_{2k-1}(n)$, Theorem 6.16 shows that the divisor functions $\sigma_\alpha(n)$ satisfy the multiplicative property in Equation (4) when $\alpha = 2k - 1$. Actually, they satisfy (4) for all real or complex α, but $\sigma_\alpha(n)$ is the nth coefficient of an entire form only when α is an odd integer ≥ 3.

EXAMPLES. The problem of determining all entire noncusp forms whose coefficients satisfy the multiplicative property (39) has been completely settled by Theorem 6.16. For the cusp forms the problem has been reduced by Theorem 6.15 to that of determining simultaneous normalized eigenforms of even weight $2k \geq 12$. We have already noted that the function $(2\pi)^{-12}\Delta(\tau)$ is the only simultaneous normalized eigenform of weight $2k = 12$. Also there is exactly one simultaneous normalized eigenform for each of the weights

$$2k = 16, 18, 20, 22, \text{ and } 26$$

since dim $M_{2k,0} = 1$ for these weights. The corresponding normalized eigenforms are given by

$$(2\pi)^{-12}\Delta(\tau) \cdot \frac{G_{2k-12}(\tau)}{2\zeta(2k-12)} = \sum_{n=1}^{\infty} \tau(n)x^n \left\{ 1 - \frac{2(2k-12)}{B_{2k-12}} \sum_{m=1}^{\infty} \sigma_{2k-13}(m)x^m \right\}.$$

We define $\tau(0) = 0$ and $\sigma_{2k-1}(0) = -B_{2k}/(4k)$. Then the coefficients $c(n)$ of these eigenforms are given by the Cauchy product

$$c(n) = -\frac{4k-24}{B_{2k-12}} \sum_{m=0}^{n} \tau(m)\sigma_{2k-13}(n-m).$$

They satisfy the multiplicative property

$$c(m)c(n) = \sum_{d|(m,n)} d^{2k-1} c\left(\frac{mn}{d^2}\right)$$

for all $m \geq 1, n \geq 1$.

6.14 Remarks on existence of simultaneous eigenforms in $M_{2k,0}$

Let $\kappa = \dim M_{2k,0}$ where $2k \geq 12$. Then we have

$$\kappa = \begin{cases} \left[\dfrac{2k}{12}\right] - 1 & \text{if } 2k \equiv 2 \pmod{12} \\[2ex] \left[\dfrac{2k}{12}\right] & \text{if } 2k \not\equiv 2 \pmod{12}. \end{cases}$$

Let $e(k)$ denote the number of linearly independent simultaneous eigenforms in $M_{2k,0}$. Clearly, $e(k) \leq \kappa$. We have shown that $e(k) = 1$ when $\kappa = 1$. Hecke showed that $e(k) = 2$ when $\kappa = 2$, and later Petersson [32] showed that $e(k) = \kappa$ in all cases. He did this by introducing an inner product (f, g) in $M_{2k,0}$ defined by the double integral

$$(f,g) = \iint_{R_\Gamma} f(\tau)\bar{g}(\tau)v^{2k-2} \, du \, dv$$

extended over the fundamental region R_Γ in the $\tau = u + iv$ plane. Relative to the Petersson inner product the Hecke operators are Hermitian, that is, they satisfy the relation

$$(T_n f, g) = (f, T_n g)$$

for any two cusp forms in $M_{2k,0}$. Therefore, by a well known theorem of linear algebra (see [2], Theorem 5.4) for each T_n there exist κ eigenforms which form an orthonormal basis for $M_{2k,0}$. These need not be simultaneous eigenforms for all the T_n. However, since the T_n commute with each other, another theorem of linear algebra (see [10], Ch. IX, Sec. 15) shows that

$M_{2k,0}$ has an orthonormal basis consisting of κ simultaneous eigenforms. Each of these can be multiplied by a constant factor to get a new basis of simultaneous normalized eigenforms. (The new basis will be orthogonal but need not be orthonormal.) Since the T_n are Hermitian, the corresponding eigenvalues are real. Details of the proofs of these statements can be found in references [32], [26], or [11].

6.15 Estimates for the Fourier coefficients of entire forms

Assume f is an entire form with Fourier expansion

$$(45) \qquad f(\tau) = \sum_{n=0}^{\infty} c(n)x^n,$$

where $x = e^{2\pi i \tau}$. Write $\tau = u + iv$ so that $x = e^{-2\pi v}e^{2\pi i u}$. For fixed $v > 0$, as u varies from 0 to 1 the point x traces out a circle $C(v)$ of radius $e^{-2\pi v}$ with center at $x = 0$. By Cauchy's residue theorem we have

$$(46) \qquad c(n) = \frac{1}{2\pi i} \int_{C(v)} \frac{f(\tau)}{x^{n+1}} \, dx = \int_0^1 f(u + iv)x^{-n} \, du.$$

We shall use this integral representation to estimate the order of magnitude of $|c(n)|$. First we consider cusp forms of weight $2k$.

Theorem 6.17. *If $f \in M_{2k,0}$ we have*

$$c(n) = O(n^k).$$

PROOF. The series in (45) converges absolutely if $|x| < 1$. Since $c(0) = 0$ we can remove a factor x and write

$$|f(\tau)| = |x| \left| \sum_{n=1}^{\infty} c(n)x^{n-1} \right| \le |x| \sum_{n=1}^{\infty} |c(n)||x|^{n-1}.$$

If τ is in R_Γ, the fundamental region of Γ, then $\tau = u + iv$ with $v \ge \sqrt{3}/2 > 1/2$, so $|x| = e^{-2\pi v} < e^{-\pi}$. Hence

$$|f(\tau)| \le A|x| = Ae^{-2\pi v}$$

where

$$A = \sum_{n=1}^{\infty} |c(n)|e^{-(n-1)\pi}.$$

This implies

$$(47) \qquad |f(\tau)|v^k \le Av^k e^{-2\pi v}.$$

Now define

$$g(\tau) = \tfrac{1}{2}|\tau - \bar{\tau}| = v$$

if $\tau \in H$. Then

$$g(A\tau) = |c\tau + d|^{-2}g(\tau)$$

if $A = \begin{pmatrix} a & b \\ c & d \end{pmatrix} \in \Gamma$, so $g^k(A\tau) = |c\tau + d|^{-2k}g^k(\tau)$. Therefore the product

$$\varphi(\tau) = |f(\tau)||g^k(\tau)| = |f(\tau)|v^k$$

is invariant under the transformations of Γ. Moreover, φ is continuous in R_Γ, and (47) shows that $\varphi(\tau) \to 0$ as $v \to +\infty$. Therefore φ is bounded in R_Γ and, since φ is invariant under Γ, φ is also bounded in H, say

$$|\varphi(\tau)| \leq M$$

for all τ in H. Therefore

$$|f(\tau)| \leq Mv^{-k}$$

for all τ in H. Using this in (46) we find

$$|c(n)| \leq \int_0^1 |f(u + iv)x^{-n}| \, du \leq Mv^{-k}|x|^{-n} = Mv^{-k}e^{2\pi nv}.$$

This holds for all $v > 0$. When $v = 1/n$ it gives us

$$|c(n)| \leq Mn^k e^{2\pi} = O(n^k).$$ □

Theorem 6.18. *If $f \in M_{2k}$ and f is not a cusp form, then*

(48) $$c(n) = O(n^{2k-1}).$$

PROOF. If $f = G_{2k}$ each coefficient $c(n)$ is of the form $\alpha\sigma_{2k-1}(n)$ where α is independent of n. Hence

$$|c(n)| \leq |\alpha|\sigma_{2k-1}(n).$$

Now

$$\sigma_{2k-1}(n) = \sum_{d|n} \left(\frac{n}{d}\right)^{2k-1} = n^{2k-1} \sum_{d|n} \frac{1}{d^{2k-1}} \leq n^{2k-1} \sum_{d=1}^{\infty} \frac{1}{d^{2k-1}} = O(n^{2k-1}),$$

so (48) holds if $f = G_{2k}$.

For a general noncusp form in M_{2k}, let $\lambda = f(i\infty)/G_{2k}(i\infty)$. Then $f - \lambda G_{2k}$ is a cusp form so

$$f = \lambda G_{2k} + g$$

where $g \in M_{2k,0}$. The Fourier coefficients of f are the sum of those of λG_{2k} and g so they have order of magnitude

$$O(n^{2k-1}) + O(n^k) = O(n^{2k-1}).$$ □

Note. For cusp forms, better estimates for the order of magnitude of the $c(n)$ have been obtained by Kloosterman, Salié, Davenport, Rankin, and Selberg (see [46]). It has been shown that

$$c(n) = O(n^{k-(1/4)+\varepsilon})$$

for every $\varepsilon > 0$, and it has been conjectured that the exponent can be further improved to $k - \frac{1}{2} + \varepsilon$. For the discriminant Δ, Ramanujan conjectured the sharper estimate

$$|\tau(p)| \le 2p^{11/2}$$

for primes p. This was recently proved by P. Deligne [7].

6.16 Modular forms and Dirichlet series

Hecke found a remarkable connection between each modular form with Fourier series

$$(49) \qquad f(\tau) = c(0) + \sum_{n=1}^{\infty} c(n)e^{2\pi i n \tau}$$

and the Dirichlet series

$$(50) \qquad \varphi(s) = \sum_{n=1}^{\infty} \frac{c(n)}{n^s}$$

formed with the same coefficients (except for $c(0)$). If $f \in M_{2k}$ then $c(n) = O(n^k)$ if f is a cusp form, and $c(n) = O(n^{2k-1})$ if f is not a cusp form. Therefore, the Dirichlet series in (50) converges absolutely for $\sigma = \operatorname{Re}(s) > k + 1$ if f is a cusp form, and for $\sigma > 2k$ if f is not a cusp form.

Theorem 6.19. *If the coefficients $c(n)$ satisfy the multiplicative property*

$$(51) \qquad c(m)c(n) = \sum_{d|(m,n)} d^{2k-1} c\left(\frac{mn}{d^2}\right)$$

the Dirichlet series will have an Euler product representation of the form

$$(52) \qquad \varphi(s) = \prod_p \frac{1}{1 - c(p)p^{-s} + p^{2k-1}p^{-2s}},$$

absolutely convergent with the Dirichlet series.

PROOF. Since the coefficients are multiplicative we have (see [4], Theorem 11.7)

$$(53) \qquad \varphi(s) = \prod_p \left\{ 1 + \sum_{n=1}^{\infty} c(p^n)p^{-ns} \right\}$$

whenever the Dirichlet series converges absolutely. Now (51) implies

$$c(p)c(p^n) = c(p^{n+1}) + p^{2k-1}c(p^{n-1})$$

for each prime p. Using this it is easy to verify the power series identity

$$(1 - c(p)x + p^{2k-1}x^2)\left(1 + \sum_{n=1}^{\infty} c(p^n)x^n\right) = 1$$

for all $|x| < 1$. Taking $x = p^{-s}$, we find that (53) reduces to (52). □

EXAMPLE. For the Ramanujan function we have the Euler product representation

$$\sum_{n=1}^{\infty} \frac{\tau(n)}{n^s} = \prod_{p} \frac{1}{1 - \tau(p)p^{-s} + p^{11-2s}}$$

for $\sigma > 7$ since $\tau(n) = O(n^6)$.

Hecke also deduced the following analytic properties of $\varphi(s)$.

Theorem 6.20. *Let $\varphi(s)$ be the function defined for $\sigma > k$ by the Dirichlet series (50) associated with a modular form $f(\tau)$ in M_k having the Fourier series (49), where k is an even integer ≥ 4. Then $\varphi(s)$ can be continued analytically beyond the line $\sigma = k$ with the following properties:*
(a) *If $c(0) = 0$, $\varphi(s)$ is an entire function of s.*
(b) *If $c(0) \neq 0$, $\varphi(s)$ is analytic for all s except for a simple pole at $s = k$ with residue*

$$\frac{(-1)^{k/2}c(0)(2\pi)^k}{\Gamma(k)}.$$

(c) *The function φ satisfies the functional equation*

$$(2\pi)^{-s}\Gamma(s)\varphi(s) = (-1)^{k/2}(2\pi)^{s-k}\Gamma(k-s)\varphi(k-s).$$

PROOF. From the integral representation for $\Gamma(s)$ we have

$$\Gamma(s)(2n\pi)^{-s} = \int_0^{\infty} e^{-2\pi n y}y^{s-1}\,dy$$

if $\sigma > 0$. Therefore if $\sigma > k$ we can multiply both members by $c(n)$ and sum on n to obtain

$$(2\pi)^{-s}\Gamma(s)\varphi(s) = \int_0^{\infty} \{f(iy) - c(0)\}y^{s-1}\,dy.$$

137

Since f is a modular form in M_k we have $f(i/y) = (iy)^k f(iy)$ so

$$(2\pi)^{-s}\Gamma(s)\varphi(s) = \int_1^\infty \{f(iy) - c(0)\}y^{s-1}\,dy + \int_0^1 \left\{(iy)^{-k}f\left(\frac{i}{y}\right) - c(0)\right\}y^{s-1}\,dy$$

$$= \int_1^\infty \{f(iy) - c(0)\}y^{s-1}\,dy + i^{-k}\int_1^\infty f(iw)w^{k-s-1}\,dw - \frac{c(0)}{s}$$

$$= \int_1^\infty \{f(iy) - c(0)\}y^{s-1}\,dy$$

$$+ (-1)^{k/2}\int_1^\infty \{f(iw) - c(0)\}w^{k-s-1}\,dw$$

$$+ (-1)^{k/2}c(0)\int_1^\infty w^{k-s-1}\,dw - \frac{c(0)}{s}$$

$$= \int_1^\infty \{f(iy) - c(0)\}(y^s + (-1)^{k/2}y^{k-s})\frac{dy}{y}$$

$$- c(0)\left(\frac{1}{s} + \frac{(-1)^{k/2}}{k-s}\right).$$

Although this last relation was proved under the assumption that $\sigma > k$, the right member is meaningful for all complex s. This gives the analytic continuation of $\varphi(s)$ beyond the line $\sigma = k$ and also verifies (a) and (b). Moreover, replacing s by $k - s$ leaves the right member unchanged except for a factor $(-1)^{k/2}$ so we also obtain (c). $\square$

Hecke also proved a converse to Theorem 6.20 to the effect that every Dirichlet series φ which satisfies a functional equation of the type in (c), together with some analytic and growth conditions, necessarily arises from a modular form in M_k. For details, see [15].

Exercises for Chapter 6

Exercises 1 through 6 deal with arithmetical functions f satisfying a relation of the form

(54)
$$f(m)f(n) = \sum_{d|(m,n)} \alpha(d)f\left(\frac{mn}{d^2}\right)$$

for all positive integers m and n, where α is a given completely multiplicative function (that is, $\alpha(1) = 1$ and $\alpha(mn) = \alpha(m)\alpha(n)$ for all m and n). An arithmetical function satisfying (54) will be called α-*multiplicative*. We write $f = 0$ if $f(n) = 0$ for all n.

1. Assume f is α-multiplicative and $f \neq 0$. Prove that $f(1) = 1$. Also prove that cf is α-multiplicative if, and only if, $c = 0$ or $c = 1$.

2. If f and g are α-multiplicative, prove that $f + g$ is α-multiplicative if, and only if, $f = 0$ or $g = 0$.

3. Let $f_1, \ldots, f_k$ be k distinct nonzero α-multiplicative functions. If a linear combination

$$f = \sum_{i=1}^{k} c_i f_i$$

is also α-multiplicative, prove that:
(a) The functions $f_1, \ldots, f_k$ are linearly independent.
(b) Either all the c_i are 0 or else exactly one of the c_i is 1 and the others are 0. Hence either $f = 0$ or $f = f_i$ for some i. In other words, linear combinations of α-multiplicative functions are never α-multiplicative except for trivial cases.

4. If f is α-multiplicative, prove that

$$\alpha(n) f(m) = \sum_{d|n} \mu(d) f(mnd) f\left(\frac{n}{d}\right).$$

5. If f is multiplicative, prove that f is α-multiplicative if, and only if,

(55) $$f(p^{k+1}) = f(p)f(p^k) - \alpha(p)f(p^{k-1})$$

for all primes p and all integers $k \geq 1$.

6. The recursion relation (55) shows that $f(p^n)$ is a polynomial in $f(p)$, say

$$f(p^n) = Q_n(f(p)).$$

The sequence $\{Q_n(x)\}$ is determined by the relations

$$Q_1(x) = x, \; Q_2(x) = x^2 - \alpha(p), \; Q_{r+1}(x) = xQ_r(x) - \alpha(p)Q_{r-1}(x) \quad \text{for } r \geq 2.$$

Show that

$$Q_n(2\alpha(p)^{1/2}x) = \alpha(p)^{n/2} U_n(x),$$

where $U_n(x)$ is the Chebyshev polynomial of the second kind, defined by the relations

$$U_1(x) = 2x, \quad U_2(x) = 4x^2 - 1, \quad U_{r+1}(x) = 2xU_r(x) - U_{r-1}(x) \quad \text{for } r \geq 1.$$

7. Let $E_{2k}(\tau) = \tfrac{1}{2} G_{2k}(\tau)/\zeta(2k)$. If $x = e^{2\pi i \tau}$ verify that the Fourier expansion of $E_{2k}(\tau)$ has the following form for $k = 2, 3, 4, 5, 6$, and 7:

$$E_4(\tau) = 1 + 240 \sum_{n=1}^{\infty} \sigma_3(n)x^n,$$

$$E_6(\tau) = 1 - 504 \sum_{n=1}^{\infty} \sigma_5(n)x^n,$$

$$E_8(\tau) = 1 + 480 \sum_{n=1}^{\infty} \sigma_7(n)x^n,$$

$$E_{10}(\tau) = 1 - 264 \sum_{n=1}^{\infty} \sigma_9(n)x^n,$$

$$E_{12}(\tau) = 1 + \frac{65520}{691} \sum_{n=1}^{\infty} \sigma_{11}(n)x^n,$$

$$E_{14}(\tau) = 1 - 24 \sum_{n=1}^{\infty} \sigma_{13}(n)x^n.$$

Derive each of the identities in Exercises 8, 9, and 10 by equating coefficients in appropriate identities involving modular forms.

8. $\sigma_7(n) = \sigma_3(n) + 120 \sum\limits_{m=1}^{n-1} \sigma_3(m)\sigma_3(n-m)$.

9. $11\sigma_9(n) = 21\sigma_5(n) - 10\sigma_3(n) + 5040 \sum\limits_{m=1}^{n-1} \sigma_3(m)\sigma_5(n-m)$.

10. $\tau(n) = \dfrac{65}{756}\sigma_{11}(n) + \dfrac{691}{756}\sigma_5(n) - \dfrac{691}{3}\sum\limits_{m=1}^{n-1}\sigma_5(m)\sigma_5(n-m)$.

   Show that this identity implies Ramanujan's congruence

$$\tau(n) \equiv \sigma_{11}(n) \pmod{691}.$$

11. Prove that the products $G_{k-12r}\Delta^r$ which occur in Theorem 6.3 are linearly independent.

12. Prove that the products $G_4{}^a G_6{}^b$ are linearly independent, where a and b are non-negative integers such that $4a + 6b = k$.

13. Show that the Dirichlet series associated with the normalized modular form

$$f(\tau) = \frac{(2k-1)!}{(2\pi i)^{2k}}\zeta(2k) + \sum_{m=1}^{\infty}\sigma_{2k-1}(m)e^{2\pi i m\tau}$$

   is $\varphi(s) = \zeta(s)\zeta(s+1-2k)$.

14. A quadratic polynomial $1 - Ax + Bx^2$ with real coefficients A and B can be factored as follows:

$$1 - Ax + Bx^2 = (1 - r_1 x)(1 - r_2 x).$$

   Prove that $r_1 = \alpha + i\beta$ and $r_2 = \gamma - i\beta$, where α, β, γ are real and $\beta(\gamma - \alpha) = 0$. Hence, if $\beta \neq 0$ the numbers r_1 and r_2 are complex conjugates.

Note. For the quadratic polynomial occurring in the proof of Theorem 6.19 we have

$$1 - c(p)x + p^{2k-1}x^2 = (1 - r_1 x)(1 - r_2 x),$$

where

$$r_1 + r_2 = c(p) \qquad \text{and} \qquad r_1 r_2 = p^{2k-1}.$$

Petersson conjectured that r_1 and r_2 are always complex conjugates. This implies

$$|r_1| = |r_2| = p^{k-1/2} \qquad \text{and} \qquad |c(p)| \leq 2p^{k-1/2}.$$

When $c(n) = \tau(n)$ this is the Ramanujan conjecture. The Petersson conjecture was proved recently by Deligne [7].

15. This exercise outlines Riemann's derivation of the functional equation

(56)
$$\pi^{-s/2}\Gamma\left(\frac{s}{2}\right)\zeta(s) = \pi^{(s-1)/2}\Gamma\left(\frac{1-s}{2}\right)\zeta(1-s)$$

from the functional equation (see Exercise 4.1)

(57)
$$\vartheta\left(\frac{-1}{\tau}\right) = (-i\tau)^{1/2}\vartheta(\tau)$$

satisfied by Jacobi's theta function

$$\vartheta(\tau) = 1 + 2\sum_{n=1}^{\infty} e^{\pi i n^2 \tau}.$$

(a) If $\sigma > 1$ prove that

$$\pi^{-s/2}\Gamma\left(\frac{s}{2}\right)n^{-s} = \int_0^\infty e^{-\pi n^2 x} x^{s/2-1}\, dx$$

and use this to derive the representation

$$\pi^{-s/2}\Gamma\left(\frac{s}{2}\right)\zeta(s) = \int_0^\infty \psi(x)x^{s/2-1}\, dx,$$

where $2\psi(x) = \vartheta(x) - 1$.

(b) Use (a) and (57) to obtain the representation

$$\pi^{-s/2}\Gamma\left(\frac{s}{2}\right)\zeta(s) = \frac{1}{s(s-1)} + \int_1^\infty (x^{s/2-1} + x^{(1-s)/2-1})\psi(x)\, dx$$

for $\sigma > 1$.

(c) Show that the equation in (b) gives the analytic continuation of $\zeta(s)$ beyond the line $\sigma = 1$ and that it also implies the functional equation (56).

7 Kronecker's theorem with applications

7.1 Approximating real numbers by rational numbers

Every irrational number θ can be approximated to any desired degree of accuracy by rational numbers. In fact, if we truncate the decimal expansion of θ after n decimal places we obtain a rational number which differs from θ by less than 10^{-n}. However, the truncated decimals might have very large denominators. For example, if

$$\theta = \pi - 3 = 0.141592653\ldots$$

the first five decimal approximations are 0.1, 0.14, 0.141, 0.1415, 0.14159. Written in the form a/b, where a and b are relatively prime integers, these rational approximations become

$$\frac{1}{10}, \quad \frac{7}{50}, \quad \frac{141}{1000}, \quad \frac{283}{2000}, \quad \frac{14159}{100,000}.$$

On the other hand, the fraction $1/7 = 0.142857\ldots$ differs from θ by less than $2/1000$ and is nearly as good as $141/1000$ for approximating θ, yet its denominator 7 is very small compared to 1000.

This example suggests the following type of question: Given a real number θ, is there a rational number h/k which is a good approximation to θ but whose denominator k is not too large?

This is, of course, a vague question because the terms "good approximation" and "not too large" are vague. Before we make the question more precise we formulate it in a slightly different way. If $\theta - h/k$ is small, then $(k\theta - h)/k$ is small. For this to be small without k being large the numerator $k\theta - h$ should be small. Therefore, we can ask the following question:

Given a real number θ and given $\varepsilon > 0$, are there integers h and k such that $|k\theta - h| < \varepsilon$?

The following theorem of Dirichlet answers this question in the affirmative.

7.2 Dirichlet's approximation theorem

Theorem 7.1. *Given any real θ and any positive integer N, there exist integers h and k with $0 < k \le N$ such that*

$$(1) \qquad\qquad |k\theta - h| < \frac{1}{N}.$$

PROOF. Let $\{x\} = x - [x]$ denote the fractional part of x. Consider the $N + 1$ real numbers

$$0, \{\theta\}, \{2\theta\}, \ldots, \{N\theta\}.$$

All these numbers lie in the half open unit interval $0 \le \{m\theta\} < 1$. Now divide the unit interval into N equal half-open subintervals of length $1/N$. Then some subinterval must contain at least two of these fractional parts, say $\{a\theta\}$ and $\{b\theta\}$, where $0 \le a < b \le N$. Hence we can write

$$(2) \qquad\qquad |\{b\theta\} - \{a\theta\}| < \frac{1}{N}.$$

But

$$\{b\theta\} - \{a\theta\} = b\theta - [b\theta] - a\theta + [a\theta] = (b - a)\theta - ([b\theta] - [a\theta]).$$

Therefore if we let

$$k = b - a \qquad \text{and} \qquad h = [b\theta] - [a\theta]$$

inequality (2) becomes

$$|k\theta - h| < \frac{1}{N}, \quad \text{with } 0 < k \le N.$$

This proves the theorem. $\qquad\qquad\qquad\qquad\qquad\qquad\qquad\qquad$ $\square$

Note. Given $\varepsilon > 0$ we can choose $N > 1/\varepsilon$ and (1) implies $|k\theta - h| < \varepsilon$.

The next theorem shows that we can choose h and k to be relatively prime.

Theorem 7.2. *Given any real θ and any positive integer N, there exist relatively prime integers h and k with $0 < k \le N$ such that*

$$|k\theta - h| < \frac{1}{N}.$$

143

PROOF. By Theorem 7.1 there is a pair h', k' with $0 < k' \leq N$ satisfying

(3)
$$\left| \theta - \frac{h'}{k'} \right| < \frac{1}{Nk'}.$$

Let $d = (h', k')$. If $d = 1$ there is nothing to prove. If $d > 1$ write $h' = hd$, $k' = kd$, where $(h, k) = 1$ and $k < k' \leq N$. Then $1/k' < 1/k$ and (3) becomes

$$\left| \theta - \frac{h}{k} \right| < \frac{1}{Nk'} < \frac{1}{Nk},$$

from which we find $|k\theta - h| < 1/N$. $\square$

Now we restate the result in a slightly weaker form which does not involve the integer N.

Theorem 7.3. *For every real θ there exist integers h and k with $k > 0$ and $(h, k) = 1$ such that*

$$\left| \theta - \frac{h}{k} \right| < \frac{1}{k^2}.$$

PROOF. In Theorem 7.2 we have $1/(Nk) \leq 1/k^2$ because $k \leq N$. $\square$

Theorem 7.4. *If θ is real, let $S(\theta)$ denote the set of all ordered pairs of integers (h, k) with $k > 0$ and $(h, k) = 1$ such that*

$$\left| \theta - \frac{h}{k} \right| < \frac{1}{k^2}.$$

Then $S(\theta)$ has the following properties:

(a) *$S(\theta)$ is nonempty.*
(b) *If θ is irrational, $S(\theta)$ is an infinite set.*
(c) *When $S(\theta)$ is infinite it contains pairs (h, k) with k arbitrarily large.*
(d) *If θ is rational, $S(\theta)$ is a finite set.*

PROOF. Part (a) is merely a restatement of Theorem 7.3. To prove (b), assume θ is irrational and assume also that $S(\theta)$ is finite. We shall obtain a contradiction. Let

$$\alpha = \min_{(h, k) \in S(\theta)} \left| \theta - \frac{h}{k} \right|.$$

Since θ is irrational, α is positive. Choose any integer $N > 1/\alpha$, for example, $N = 1 + [1/\alpha]$. Then $1/N < \alpha$. Applying Theorem 7.2 with this N we obtain a pair of integers h and k with $(h, k) = 1$ and $0 < k \leq N$ such that

$$\left| \theta - \frac{h}{k} \right| < \frac{1}{kN}.$$

Now $1/(kN) \leq 1/k^2$ so the pair $(h, k) \in S(\theta)$. But we also have

$$\frac{1}{kN} \leq \frac{1}{N} < \alpha, \qquad \text{so} \qquad \left| \theta - \frac{h}{k} \right| < \alpha,$$

contradicting the definition of α. This shows that $S(\theta)$ cannot be finite if θ is irrational.

To prove (c) assume that all pairs (h, k) in $S(\theta)$ have $k \leq M$ for some M. We will show that this leads to a contradiction by showing that the number of choices for h is also bounded. If $(h, k) \in S(\theta)$ we have

$$|k\theta - h| < \frac{1}{k} \leq 1,$$

so

$$|h| = |h - k\theta + k\theta| \leq |h - k\theta| + |k\theta| < 1 + |k\theta| \leq 1 + M|\theta|.$$

Therefore the number of choices for h is bounded, contradicting the fact that $S(\theta)$ is infinite.

To prove (d), assume θ is rational, say $\theta = a/b$, where $(a, b) = 1$ and $b > 0$. Then the pair $(a, b) \in S(\theta)$ because $\theta - a/b = 0$. Now we assume that $S(\theta)$ is an infinite set and obtain a contradiction. If $S(\theta)$ is infinite then by part (c) there is a pair (h, k) in $S(\theta)$ with $k > b$. For this pair we have

$$0 < \left| \frac{a}{b} - \frac{h}{k} \right| < \frac{1}{k^2},$$

from which we find $0 < |ak - bh| < b/k < 1$. This is a contradiction because $ak - bh$ is an integer. $\qquad \square$

Theorem 7.4 shows that a real number θ is irrational if, and only if, there are infinitely many rational numbers h/k with $(h, k) = 1$ and $k > 0$ such that

$$\left| \theta - \frac{h}{k} \right| < \frac{1}{k^2}.$$

This inequality can be improved. It is easy to show that the numerator 1 can be replaced by $\frac{1}{2}$ (see Exercise 7.4). Hurwitz replaced $\frac{1}{2}$ by a smaller constant. He proved that θ is irrational if, and only if, there exist infinitely many rational numbers h/k with $(h, k) = 1$ and $k > 0$ such that

$$\left| \theta - \frac{h}{k} \right| < \frac{1}{\sqrt{5}k^2}.$$

Moreover, the result is false if $1/\sqrt{5}$ is replaced by any smaller constant. (See Exercise 7.5.) We shall not prove Hurwitz's theorem. Instead, we prove a theorem of Liouville which shows that the denominator k^2 cannot be replaced by k^3 or any higher power.

7.3 Liouville's approximation theorem

Theorem 7.5. *Let θ be a real algebraic number of degree $n \geq 2$. Then there is a positive constant $C(\theta)$, depending only on θ, such that for all integers h and k with $k > 0$ we have*

$$(4) \qquad \left| \theta - \frac{h}{k} \right| > \frac{C(\theta)}{k^n}.$$

PROOF. Since θ is algebraic of degree n, θ is a zero of some polynomial $f(x)$ of degree n with integer coefficients, say

$$f(x) = \sum_{r=0}^{n} a_r x^r,$$

where $f(x)$ is irreducible over the rational field. Since $f(x)$ is irreducible it has no rational roots so $f(h/k) \neq 0$ for every rational h/k.

Now we use the mean value theorem of differential calculus to write

$$(5) \qquad f\left(\frac{h}{k}\right) = f\left(\frac{h}{k}\right) - f(\theta) = f'(\xi)\left(\frac{h}{k} - \theta\right),$$

where ξ lies between θ and h/k. We will deduce (4) from (5) by getting an upper bound for $|f'(\xi)|$ and a lower bound for $|f(h/k)|$. We have

$$f\left(\frac{h}{k}\right) = \sum_{r=0}^{n} a_r \left(\frac{h}{k}\right)^r = \frac{N}{k^n}$$

where N is a nonzero integer. Therefore

$$(6) \qquad \left| f\left(\frac{h}{k}\right) \right| \geq \frac{1}{k^n},$$

which is the required lower bound. To get an upper bound for $|f'(\xi)|$ we let

$$d = \left| \theta - \frac{h}{k} \right|.$$

If $d > 1$ then (4) holds with $C(\theta) = 1$, so we can assume that $d < 1$. (We cannot have $d = 1$ since θ is irrational.) Since ξ lies between θ and h/k and $d < 1$ we have $|\xi - \theta| < 1$ so

$$|\xi| = |\theta + \xi - \theta| \leq |\theta| + |\xi - \theta| < |\theta| + 1.$$

Hence

$$|f'(\xi)| \leq A(\theta) < 1 + A(\theta),$$

where $A(\theta)$ denote the maximum value of $|f'(x)|$ in the interval $|x| \leq |\theta| + 1$. Using this upper bound for $|f'(\xi)|$ in (5) together with the lower bound in (6) we obtain (4) with $C(\theta) = 1/(1 + A(\theta))$. $\qquad \square$

A real number which is not algebraic is called *transcendental*. A simple counting argument shows that transcendental numbers exist. In fact, the set of all real algebraic numbers is countable, but the set of all real numbers is uncountable, so the transcendental numbers not only exist but they form an uncountable set.

It is usually difficult to show that some particular number such as e or π is transcendental. Liouville's theorem can be used to show that irrational numbers that are sufficiently well approximated by rationals are necessarily transcendental. Such numbers are called Liouville numbers and are defined as follows.

Definition. A real number θ is called a *Liouville number* if for every integer $r \geq 1$ there exist integers h_r and k_r with $k_r > 0$ such that

(7)
$$0 < \left| \theta - \frac{h_r}{k_r} \right| < \frac{1}{k_r{}^r}.$$

Theorem 7.6. *Every Liouville number is transcendental.*

PROOF. If a Liouville number θ were algebraic of degree n it would satisfy both inequality (7) and

$$\left| \theta - \frac{h_r}{k_r} \right| > \frac{C(\theta)}{k_r{}^n}$$

for every $r \geq 1$, where $C(\theta)$ is the constant in Theorem 7.5. Therefore

$$0 < \frac{C(\theta)}{k_r{}^n} < \frac{1}{k_r{}^r}, \quad \text{or} \quad 0 < C(\theta) < \frac{1}{k_r{}^{r-n}}.$$

The last inequality gives a contradiction if r is sufficiently large.   □

EXAMPLE. The number

$$\theta = \sum_{m=1}^{\infty} \frac{1}{10^{m!}}$$

is a Liouville number and hence is transcendental. In fact, for each $r \geq 1$ we can take $k_r = 10^{r!}$ and

$$h_r = k_r \sum_{m=1}^{r} \frac{1}{10^{m!}}.$$

Then we have

$$0 < \theta - \frac{h_r}{k_r} = \sum_{m=r+1}^{\infty} \frac{1}{10^{m!}} \leq \frac{1}{10^{(r+1)!}} \sum_{m=0}^{\infty} \frac{1}{10^m}$$

$$= \frac{10/9}{10^{(r+1)!}} = \frac{1}{k_r{}^r} \frac{10/9}{10^{r!}} < \frac{1}{k_r{}^r}$$

so (7) is satisfied.

147

Note. The same argument shows that $\sum_{m=1}^{\infty} a_m 10^{-m!}$ is transcendental if $a_m = 0$ or 1 and $a_m = 1$ for infinitely many m.

We turn now to a generalization of Dirichlet's theorem due to Kronecker.

7.4 Kronecker's approximation theorem: the one-dimensional case

Dirichlet's theorem tells us that for any real θ and every $\varepsilon > 0$ there exist integers x and y, not both 0, such that

$$|\theta x + y| < \varepsilon.$$

In other words, the linear form $\theta x + y$ can be made arbitrarily close to 0 by a suitable choice of integers x and y. If θ is rational this is trivial because we can make $\theta x + y = 0$, so the result is significant only if θ is irrational. Kronecker proved a much stronger result. He showed that if θ is irrational the linear form $\theta x + y$ can be made arbitrarily close to any prescribed real number α. We prove this result first for α in the unit interval. As in the proof of Dirichlet's theorem we make use of the fractional parts $\{n\theta\} = n\theta - [n\theta]$.

Theorem 7.7. *If θ is a given irrational number the sequence of numbers $\{n\theta\}$ is dense in the unit interval. That is, given any α, $0 \le \alpha \le 1$, and given any $\varepsilon > 0$, there exists a positive integer k such that*

$$|\{k\theta\} - \alpha| < \varepsilon.$$

Hence, if $h = [k\theta]$ we have $|k\theta - h - \alpha| < \varepsilon$.

Note. This shows that the linear form $\theta x + y$ can be made arbitrarily close to α by a suitable choice of integers x and y.

PROOF. First we note that $\{n\theta\} \ne \{m\theta\}$ if $m \ne n$ because θ is irrational. Also, there is no loss of generality if we assume $0 < \theta < 1$ since $n\theta = n[\theta] + n\{\theta\}$ and $\{n\theta\} = \{n\{\theta\}\}$.

Let $\varepsilon > 0$ be given and choose any α, $0 \le \alpha \le 1$. By Dirichlet's approximation theorem there exist integers h and k such that $|k\theta - h| < \varepsilon$. Now either $k\theta > h$ or $k\theta < h$. Suppose that $k\theta > h$, so that $0 < \{k\theta\} < \varepsilon$. (The argument is similar if $k\theta < h$.) Now consider the following subsequence of the given sequence $\{n\theta\}$:

$$\{k\theta\}, \{2k\theta\}, \{3k\theta\}, \ldots.$$

We will show that the early terms of this sequence are increasing. We have

$$k\theta = [k\theta] + \{k\theta\}, \quad \text{so } mk\theta = m[k\theta] + m\{k\theta\}.$$

Hence

$$\{mk\theta\} = m\{k\theta\} \quad \text{if, and only if, } \{k\theta\} < \frac{1}{m}.$$

Now choose the largest integer N which satisfies $\{k\theta\} < 1/N$. Then we have

$$\frac{1}{N + 1} < \{k\theta\} < \frac{1}{N}.$$

Therefore $\{mk\theta\} = m\{k\theta\}$ for $m = 1, 2, \ldots, N$, so the N numbers

$$\{k\theta\}, \{2k\theta\}, \ldots, \{Nk\theta\}$$

form an increasing equally-spaced chain running from left to right in the interval $(0, 1)$. The last member of this chain (by the definition of N) satisfies the inequality

$$\frac{N}{N + 1} < \{Nk\theta\} < 1,$$

or

$$1 - \frac{1}{N + 1} < \{Nk\theta\} < 1.$$

Thus $\{Nk\theta\}$ differs from 1 by less than $1/(N + 1) < \{k\theta\} < \varepsilon$. Therefore the first N members of the subsequence $\{nk\theta\}$ subdivide the unit interval into subintervals of length $< \varepsilon$. Since α lies in one of these subintervals, the theorem is proved. □

The next theorem removes the restriction $0 \le \alpha \le 1$.

Theorem 7.8. *Given any real α, any irrational θ, and any $\varepsilon > 0$, there exist integers h and k with $k > 0$ such that*

$$|k\theta - h - \alpha| < \varepsilon.$$

PROOF. Write $\alpha = [\alpha] + \{\alpha\}$. By Theorem 7.7 there exists $k > 0$ such that $|\{k\theta\} - \{\alpha\}| < \varepsilon$. Hence

$$|k\theta - [k\theta] - (\alpha - [\alpha])| < \varepsilon$$

or

$$|k\theta - ([k\theta] - [\alpha]) - \alpha| < \varepsilon.$$

Now take $h = [k\theta] - [\alpha]$ to complete the proof. □

7.5 Extension of Kronecker's theorem to simultaneous approximation

We turn now to a problem of simultaneous approximation. Given n irrational numbers $\theta_1, \theta_2, \ldots, \theta_n$, and n real numbers $\alpha_1, \alpha_2, \ldots, \alpha_n$, and given $\varepsilon > 0$, we seek integers $h_1, h_2, \ldots, h_n$ and k such that

$$|k\theta_i - h_i - \alpha_i| < \varepsilon \quad \text{for } i = 1, 2, \ldots, n.$$

It turns out that this problem cannot always be solved as stated. For example, suppose we start with two irrational numbers, say θ_1 and $2\theta_1$, and two real numbers α_1 and α_2, and suppose there exist integers h_1, h_2 and k such that

$$|k\theta_1 - h_1 - \alpha_1| < \varepsilon$$

and

$$|2k\theta_1 - h_2 - \alpha_2| < \varepsilon.$$

Multiply the first inequality by 2 and subtract from the second to obtain

$$|2h_1 - h_2 + 2\alpha_1 - \alpha_2| < 3\varepsilon.$$

Since ε, α_1 and α_2 are arbitrary and h_1, h_2 are integers, this inequality cannot in general be satisfied. The difficulty with this example is that θ_1 and $2\theta_1$ are linearly dependent and we were able to eliminate θ_1 from the two inequalities. Kronecker showed that the problem of simultaneous approximation can always be solved if $\theta_1, \ldots, \theta_n$ are linearly independent over the integers; that is, if

$$\sum_{i=1}^{n} c_i \theta_i = 0$$

with integer multipliers $c_1, \ldots, c_n$ implies $c_1 = \cdots = c_n = 0$. This restriction is compensated for, in part, by removing the restriction that the θ_i be irrational. First we prove what appears to be a less general result.

Theorem 7.9 (First form of Kronecker's theorem). *If $\alpha_1, \ldots, \alpha_n$ are arbitrary real numbers, if $\theta_1, \ldots, \theta_n$ are linearly independent real numbers, and if $\varepsilon > 0$ is arbitrary, then there exists a real number t and integers $h_1, \ldots, h_n$ such that*

$$|t\theta_i - h_i - \alpha_i| < \varepsilon \quad for \ i = 1, 2, \ldots, n.$$

Note. The theorem exhibits a real number t, whereas we asked for an integer k. Later we show that it is possible to replace t by an integer k, but in most applications of the theorem the real t suffices.

The proof of Theorem 7.9 makes use of three lemmas.

Lemma 1. *Let $\{\lambda_n\}$ be a sequence of distinct real numbers. For each real t and arbitrary complex numbers $c_0, \ldots, c_N$ define*

$$f(t) = \sum_{r=0}^{N} c_r e^{it\lambda_r}.$$

Then for each k we have

$$c_k = \lim_{T \to \infty} \frac{1}{T} \int_0^T f(t) e^{-it\lambda_k} \, dt.$$

PROOF. The definition of $f(t)$ gives us

$$f(t)e^{-it\lambda_k} = \sum_{r=0}^{N} c_r e^{i(\lambda_r - \lambda_k)t}.$$

Hence

$$\int_0^T f(t)e^{-it\lambda_k}\, dt = \sum_{\substack{r=0 \\ r \neq k}}^{N} c_r \int_0^T e^{i(\lambda_r - \lambda_k)t}\, dt + c_k T,$$

from which we find

$$\frac{1}{T}\int_0^T f(t)e^{-it\lambda_k}\, dt = c_k + \sum_{\substack{r=0 \\ r \neq k}}^{N} c_r \frac{e^{i(\lambda_r - \lambda_k)T} - 1}{i(\lambda_r - \lambda_k)T}.$$

Now let $T \to \infty$ to obtain the lemma. $\qquad \square$

Lemma 2. *If t is real, let*

(8) $$F(t) = 1 + \sum_{r=1}^{n} e^{2\pi i(t\theta_r - \alpha_r)},$$

where $\alpha_1, \ldots, \alpha_n$ and $\theta_1, \ldots, \theta_n$ are arbitrary real numbers. Let

$$L = \sup_{-\infty < t < +\infty} |F(t)|.$$

Then the following two statements are equivalent:

(a) *For every $\varepsilon > 0$ there exists a real t and integers $h_1, \ldots, h_n$ such that*

$$|t\theta_r - \alpha_r - h_r| < \varepsilon \quad for\ r = 1, 2, \ldots, n.$$

(b) $L = n + 1$.

PROOF. The idea of the proof is fairly simple. Each term of the sum in (8) has absolute value 1 so $|F(t)| \leq n + 1$. If (a) holds then each number $t\theta_r - \alpha_r$ is nearly an integer hence each exponential in (8) is nearly 1 so $|F(t)|$ is nearly $n + 1$. Conversely, if (b) holds then $|F(t)|$ is nearly $n + 1$ for some t hence every term in (8) must be nearly 1 since no term has absolute value greater than 1. Therefore each number $t\theta_r - \alpha_r$ is nearly an integer so (a) holds. Now we transform this idea into a rigorous proof.

First we show that (a) implies (b). If (a) holds take $\varepsilon = 1/(2\pi k)$, where $k \geq 1$, and let t_k be the corresponding value of t given by (a). Then $2\pi(t_k\theta_r - \alpha_r)$ differs from an integer multiple of 2π by less than $1/k$ so

$$\cos 2\pi(t_k\theta_r - \alpha_r) \geq \cos \frac{1}{k}.$$

Hence

$$|F(t_k)| \geq 1 + \sum_{r=1}^{n} \cos 2\pi(t_k\theta_r - \alpha_r) \geq 1 + n \cos \frac{1}{k}$$

and therefore $L \geq |F(t_k)| \geq 1 + n \cos(1/k)$. Letting $k \to \infty$ we find $L \geq n + 1$. Since $L \leq n + 1$ this proves (b).

Now we assume (a) is false and show that (b) is also false. If (a) is false there exists an $\varepsilon > 0$ such that for all integers $h_1, \ldots, h_n$ and all real t there is a k, $1 \leq k \leq n$, such that

(9)
$$|t\theta_k - \alpha_k - h_k| \geq \frac{\varepsilon}{2\pi}.$$

(We can also assume that $\varepsilon \leq \pi/4$ because if (a) is false for ε it is also false for every smaller ε.) Let $x_r = t\theta_r - \alpha_r - h_r$. Then (9) implies $|2\pi x_k| \geq \varepsilon$ so the point $1 + e^{2\pi i x_k}$ lies on the circle of radius 1 about 1 but outside the shaded sector shown in Figure 7.1.

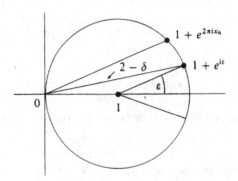

Figure 7.1

Now $|1 + e^{i\varepsilon}| < 2$ so $|1 + e^{i\varepsilon}| = 2 - \delta$ for some $\delta > 0$. Hence
$$|1 + e^{2\pi i x_k}| \leq |1 + e^{i\varepsilon}| = 2 - \delta,$$

so
$$|F(t)| = \left| 1 + \sum_{r=1}^{n} e^{2\pi i x_r} \right| \leq |1 + e^{2\pi i x_k}| + \sum_{\substack{r=1 \\ r \neq k}}^{n} |e^{2\pi i x_r}|$$

$$\leq (2 - \delta) + (n - 1) = n + 1 - \delta.$$

Since this is true for all t we must have $L \leq n + 1 - \delta < n + 1$, contradicting (b). $\qquad \square$

Lemma 3. *Let* $g = g(x_1, \ldots, x_n)$ *be the polynomial in n variables given by*

$$g = 1 + x_1 + x_2 + \cdots + x_n,$$

and write

(10)
$$g^p = 1 + \sum a_{r_1, \ldots, r_n} x_1^{r_1} \cdots x_n^{r_n},$$

where p is a positive integer. Then the coefficients $a_{r_1,\ldots,r_n}$ are positive integers such that

(11)
$$1 + \sum a_{r_1,\ldots,r_n} = (1 + n)^p,$$

and the number of terms in (10) is at most $(p + 1)^n$.

PROOF. Since $1 + \sum a_{r_1,\ldots,r_n} = g^p(1, 1, \ldots, 1) = (1 + n)^p$ this proves (11). Let $1 + N$ be the number of terms in (10). We shall prove that

(12)
$$1 + N \leq (p + 1)^n$$

by induction on n. For $n = 1$ we have

$$(1 + x_1)^p = 1 + \binom{p}{1}x_1 + \binom{p}{2}x_1^2 + \cdots + x_1^p$$

and the sum on the right has exactly $p + 1$ terms. Thus (12) holds for $n = 1$. If $n > 1$ we have

$$g_p = \{(1 + x_1 + \cdots + x_{n-1}) + x_n\}^p$$

$$= (1 + x_1 + \cdots + x_{n-1})^p + \binom{p}{1}(1 + \cdots + x_{n-1})^{p-1}x_n + \cdots + x_n^p,$$

so if there are at most $(p + 1)^{n-1}$ terms in each group on the right there will be at most $(p + 1)^n$ terms altogether. This proves (12) by induction. $\square$

PROOF OF KRONECKER'S THEOREM. Choosing $F(t)$ as in Lemma 2 we have

$$F(t) = 1 + \sum_{r=1}^{n} e^{2\pi i(t\theta_r - \alpha_r)}.$$

By Lemma 2, to prove Kronecker's theorem it suffices to prove that

$$L = \sup_{-\infty < t < +\infty} |F(t)| = n + 1.$$

The pth power of $F(t)$ is a sum of the type discussed in Lemma 1,

(13)
$$f(t) = F^p(t) = 1 + \sum_{r=1}^{N} c_r e^{it\lambda_r},$$

with $\lambda_0 = 1$ and λ_r replaced by $2\pi(r_1\theta_1 + \cdots + r_n\theta_n)$ if $r \geq 1$. The numbers λ_r are distinct because the θ_i are linearly independent over the integers. The coefficients c_r in (13) are the integers $a_{r_1,\ldots,r_n}$ of Lemma 3 multiplied by a factor of absolute value 1. Hence (11) implies

(14)
$$1 + \sum_{r=1}^{N} |c_r| = 1 + \sum a_{r_1,\ldots,r_n} = (1 + n)^p.$$

By Lemma 1 we have

(15)
$$c_r = \lim_{T \to \infty} \frac{1}{T} \int_0^T F^p(t) e^{-it\lambda_r} \, dt.$$

Now $|F(t)| \leq L$ so $|F^p(t)| \leq L^p$ for all t, hence

$$\left| \frac{1}{T} \int_0^T F^p(t) e^{-it\lambda_r} \, dt \right| \leq \frac{1}{T} \int_0^T L^p \, dt = L^p.$$

Hence (15) implies $|c_r| \leq L^p$ for each r, and (14) gives us

$$(1 + n)^p \leq (N + 1)L^p \leq (p + 1)^n L^p$$

by Lemma 3. Therefore

$$\frac{n + 1}{L} \leq (p + 1)^{n/p}$$

from which we find

$$\log\left(\frac{n + 1}{L}\right) \leq \frac{n}{p} \log(p + 1).$$

Now let $p \to \infty$. The last inequality becomes $\log[(n + 1)/L] \leq 0$, so $L \geq n + 1$. But $L \leq n + 1$ hence $L = n + 1$, and this proves Kronecker's theorem. $\qquad\square$

The next version of Kronecker's theorem replaces the real number t by an integer k.

Theorem 7.10 (Second form of Kronecker's theorem). *If $\alpha_1, \ldots, \alpha_n$ are arbitrary real numbers, if $\theta_1, \ldots, \theta_n, 1$ are linearly independent real numbers, and if $\varepsilon > 0$ is given, then there exists an integer k and integers $m_1, \ldots, m_n$ such that*

$$|k\theta_i - m_i - \alpha_i| < \varepsilon \quad for \ i = 1, 2, \ldots, n.$$

PROOF. We apply the first form of Kronecker's theorem to the system $\alpha_1, \ldots, \alpha_n, 0$ and $\{\theta_1\}, \{\theta_2\}, \ldots, \{\theta_n\}, 1$, with $\varepsilon/2$ instead of ε, where $\varepsilon < 1$. Then there exists a real t and integers $h_1, \ldots, h_{n+1}$ such that

$$|t\{\theta_i\} - h_i - \alpha_i| < \frac{\varepsilon}{2} \quad for \ i = 1, 2, \ldots, n$$

and

(16) $$|t - h_{n+1}| < \frac{\varepsilon}{2}.$$

The last inequality shows that t is nearly equal to the integer h_{n+1}. Take $k = h_{n+1}$. Then (16) implies

$$|k\{\theta_i\} - h_i - \alpha_i| = |t\{\theta_i\} - h_i - \alpha_i + (k - t)\{\theta_i\}|$$
$$\leq |t\{\theta_i\} - h_i - \alpha_i| + |k - t| < \varepsilon.$$

Hence, writing $\{\theta_i\} = \theta_i - [\theta_i]$, we obtain

$$|k(\theta_i - [\theta_i]) - h_i - \alpha_i| < \varepsilon$$

or, what is the same thing,

$$|k\theta_i - (h_i + k[\theta_i]) - \alpha_i| < \varepsilon.$$

Putting $m_i = h_i + k[\theta_i]$ we obtain the theorem. $\qquad\square$

7.6 Applications to the Riemann zeta function

With the help of Kronecker's theorem we can determine the least upper bound and greatest lower bound of $|\zeta(\sigma + it)|$ on any fixed line $\sigma =$ constant, $\sigma > 1$.

Definition. For fixed σ, we define

$$m(\sigma) = \inf_t |\zeta(\sigma + it)| \qquad \text{and} \qquad M(\sigma) = \sup_t |\zeta(\sigma + it)|,$$

where the infimum and supremum are taken over all real t.

Theorem 7.11. *For each fixed $\sigma > 1$ we have*

$$M(\sigma) = \zeta(\sigma) \qquad \text{and} \qquad m(\sigma) = \frac{\zeta(2\sigma)}{\zeta(\sigma)}.$$

PROOF. For $\sigma > 1$ we have $|\zeta(\sigma + it)| \leq \zeta(\sigma)$ so $M(\sigma) = \zeta(\sigma)$, the supremum being attained on the real axis. To obtain the result for $m(\sigma)$ we estimate the reciprocal $|1/\zeta(s)|$. For $\sigma > 1$ we have

$$(17) \qquad \left|\frac{1}{\zeta(s)}\right| = \prod_p |1 - p^{-s}| \leq \prod_p (1 + p^{-\sigma}) = \frac{\zeta(\sigma)}{\zeta(2\sigma)}.$$

Hence $|\zeta(s)| \geq \zeta(2\sigma)/\zeta(\sigma)$ so $m(\sigma) \geq \zeta(2\sigma)/\zeta(\sigma)$.

Now we wish to prove the reverse inequality $m(\sigma) \leq \zeta(2\sigma)/\zeta(\sigma)$. The idea is to show that the inequality

$$|1 - p^{-s}| \leq 1 + p^{-\sigma}$$

used in (17) is very nearly an equality for certain values of t. Now

$$1 - p^{-s} = 1 - p^{-\sigma - it} = 1 - p^{-\sigma} e^{-it \log p} = 1 + p^{-\sigma} e^{i(-t \log p - \pi)},$$

so we need to show that $-t \log p - \pi$ is nearly an even multiple of 2π for certain values of t. For this we invoke Kronecker's theorem. Of course, there are infinitely many terms in the Euler product for $1/\zeta(s)$ and we cannot expect to make $-t \log p - \pi$ nearly an even multiple of 2π for *all* primes p. But we will be able to do this for enough primes to obtain the desired inequality.

155

Choose any ε, $0 < \varepsilon < \pi/2$, and choose any integer $n \geq 1$. We apply Kronecker's theorem to the numbers

$$\theta_k = \frac{-1}{2\pi} \log p_k, \quad k = 1, 2, \ldots, n,$$

where $p_1, \ldots, p_n$ are the first n primes. The θ_i are linearly independent because

$$\sum_{i=1}^{n} a_i \log p_i = 0 \quad \text{implies} \quad \log(p_1^{a_1} \cdots p_n^{a_n}) = 0$$

so $p_1^{a_1} \cdots p_n^{a_n} = 1$ hence each $a_i = 0$. We also take $\alpha_1 = \alpha_2 = \cdots = \alpha_n = \frac{1}{2}$. Then by Theorem 7.9 there is a real t and integers $h_1, \ldots, h_n$ such that $|t\theta_k - \alpha_k - h_k| < \varepsilon/(2\pi)$, which means

(18) $$|-t \log p_k - \pi - 2\pi h_k| < \varepsilon.$$

For this t we have

$$1 - p_k^{-s} = 1 - p_k^{-\sigma} e^{-it \log p_k} = 1 + p_k^{-\sigma} e^{i(-t \log p_k - \pi)}$$
$$= 1 + p_k^{-\sigma} \cos(-t \log p_k - \pi) + i p_k^{-\sigma} \sin(-t \log p_k - \pi),$$

so

$$|1 - p_k^{-s}| \geq 1 + p_k^{-\sigma} \cos(-t \log p_k - \pi).$$

But (18) implies

$$\cos|-t \log p_k - \pi| = \cos|-t \log p_k - \pi - 2\pi h_k| > \cos \varepsilon,$$

since the cosine function decreases in the interval $[0, \pi/2]$. Hence

$$|1 - p_k^{-s}| > 1 + p_k^{-\sigma} \cos \varepsilon.$$

Now consider any partial product of the Euler product for $1/\zeta(s)$. For a given ε and n there exists a real t (depending on ε and on n) such that

(19) $$\left| \prod_{k=1}^{n} (1 - p_k^{-s}) \right| = \prod_{k=1}^{n} |1 - p_k^{-s}| > \prod_{k=1}^{n} (1 + p_k^{-\sigma} \cos \varepsilon).$$

Now

$$\frac{1}{|\zeta(s)|} = \prod_{k=1}^{\infty} |1 - p_k^{-s}|$$

and hence, by the Cauchy condition for convergent products, there is an n_0 such that $n \geq n_0$ implies

$$\left| \prod_{k=n+1}^{\infty} |1 - p_k^{-s}| - 1 \right| < \varepsilon$$

or

$$1 - \varepsilon < \prod_{k=n+1}^{\infty} |1 - p_k^{-s}| < 1 + \varepsilon.$$

Using (19) with $n \geq n_0$ we have

$$\frac{1}{|\zeta(s)|} = \prod_{k=1}^{n} |1 - p_k^{-s}| \prod_{k=n+1}^{\infty} |1 - p_k^{-s}| > (1 - \varepsilon) \prod_{k=1}^{n} (1 + p_k^{-\sigma} \cos \varepsilon).$$

This holds for $n \geq n_0$ and a certain t depending on n and on ε. Hence

$$\frac{1}{m(\sigma)} = \frac{1}{\inf_t |\zeta(\sigma + it)|} = \sup_t \frac{1}{|\zeta(\sigma + it)|} \geq (1 - \varepsilon) \prod_{k=1}^{n} (1 + p_k^{-\sigma} \cos \varepsilon).$$

Letting $n \to \infty$ we find

$$\frac{1}{m(\sigma)} \geq (1 - \varepsilon) \prod_{k=1}^{\infty} (1 + p_k^{-\sigma} \cos \varepsilon).$$

We will show in a moment that the last product converges uniformly for $0 \leq \varepsilon \leq \pi/2$. Therefore we can let $\varepsilon \to 0$ and pass to the limit term by term to obtain

$$\frac{1}{m(\sigma)} \geq \prod_{k=1}^{\infty} (1 + p_k^{-\sigma}) = \frac{\zeta(\sigma)}{\zeta(2\sigma)}.$$

This gives the desired inequality $m(\sigma) \leq \zeta(2\sigma)/\zeta(\sigma)$.

To prove the uniform convergence of the product, we use the fact that a product $\prod (1 + f_n(z))$ converges uniformly on a set if, and only if, the series $\sum f_n(z)$ converges uniformly on this set. Therefore we consider the series $\sum p_k^{-\sigma} \cos \varepsilon$. But this is dominated by $\sum p_k^{-\sigma} \leq \sum n^{-\sigma} = \zeta(\sigma)$ so the convergence is uniform in the interval $0 \leq \varepsilon \leq \pi/2$, and the proof is complete. $\square$

7.7 Applications to periodic functions

We say that n complex numbers $\omega_1, \omega_2, \ldots, \omega_n$ are linearly independent over the integers if no linear combination

$$a_1 \omega_1 + a_2 \omega_2 + \cdots + a_n \omega_n$$

with integers coefficients is 0 except when $a_1 = a_2 = \cdots = a_n = 0$. Otherwise the numbers $\omega_1, \ldots, \omega_n$ are called linearly dependent over the integers.

Elliptic functions are meromorphic functions with two linearly independent periods. In this section we use Kronecker's theorem to show that there are no meromorphic functions with three linearly independent periods except for constant functions.

Theorem 7.12. *Let ω_1 and ω_2 be periods of f such that the ratio ω_2/ω_1 is real and irrational. Then f has arbitrarily small nonzero periods. That is, given $\varepsilon > 0$ there is a period ω such that $0 < |\omega| < \varepsilon$.*

PROOF. We apply Dirichlet's approximation theorem. Let $\theta = \omega_2/\omega_1$. Since θ is irrational, given any $\varepsilon > 0$ there exist integers h and k with $k > 0$ such that

$$|k\theta - h| < \frac{\varepsilon}{|\omega_1|}.$$

Multiplying by $|\omega_1|$ we find

$$|k\omega_2 - h\omega_1| < \varepsilon.$$

But $\omega = k\omega_2 - h\omega_1$ is a period of f with $|\omega| < \varepsilon$. Also, $\omega \neq 0$ since ω_2/ω_1 is irrational. $\qquad\square$

Theorem 7.13. *If f has three periods $\omega_1, \omega_2, \omega_3$ which are linearly independent over the integers, then f has arbitrarily small nonzero periods.*

PROOF. Suppose first that ω_2/ω_1 is real. If ω_2/ω_1 is rational then ω_1 and ω_2 are linearly dependent over the integers, hence $\omega_1, \omega_2, \omega_3$ are also dependent, contradicting the hypothesis. If ω_2/ω_1 is irrational, then f has arbitrarily small nonzero periods by Theorem 7.12.

Now suppose ω_2/ω_1 is not real. Geometrically, this means that ω_1 and ω_2 are not collinear with the origin. Hence ω_3 can be expressed as a linear combination of ω_1 and ω_2 with real coefficients, say

$$\omega_3 = \alpha\omega_1 + \beta\omega_2, \quad \text{where } \alpha \text{ and } \beta \text{ are real.}$$

Now we consider three cases:

(a) Both α and β rational.
(b) One of α, β rational, the other irrational.
(c) Both α and β irrational.

Case (a) implies $\omega_1, \omega_2, \omega_3$ are dependent over the integers, contradicting the hypothesis.

For case (b), assume α is rational, say $\alpha = a/b$, and β is irrational. Then we have

$$\omega_3 = \frac{a}{b}\omega_1 + \beta\omega_2, \quad \text{so} \quad b\omega_3 - a\omega_1 = \beta(b\omega_2).$$

This gives us two periods $b\omega_3 - a\omega_1$ and $b\omega_2$ with irrational ratio, hence f has arbitrarily small periods. The same argument works, of course, if β is rational and α is irrational.

Now consider case (c), both α and β irrational. Here we consider two subcases.

(c_1) Assume α and β are linearly dependent over the integers. Then there exist integers a and b, not both zero, such that $a\alpha + b\beta = 0$. By symmetry, we can assume that $b \neq 0$. Then $\beta = -a\alpha/b$ and

$$\omega_3 = \alpha\omega_1 - \frac{a}{b}\alpha\omega_2, \quad \text{so} \quad b\omega_3 = \alpha(b\omega_1 - a\omega_2).$$

Again we have two periods $b\omega_3$ and $b\omega_1 - a\omega_2$ with irrational ratio, so f has arbitrarily small nonzero periods.

(c_2) Assume α and β are linearly independent over the integers. Then by Kronecker's theorem, given any $\varepsilon > 0$ there exist integers h_1, h_2 and k

such that

$$|k\alpha - h_1| < \frac{\varepsilon}{1 + |\omega_1| + |\omega_2|}, \qquad |k\beta - h_2| < \frac{\varepsilon}{1 + |\omega_1| + |\omega_2|}.$$

Multiply these inequalities by $|\omega_1|, |\omega_2|$, respectively, to get

$$|k\alpha\omega_1 - h_1\omega_1| < \frac{\varepsilon|\omega_1|}{1 + |\omega_1| + |\omega_2|}, \qquad |k\beta\omega_2 - h_2\omega_2| < \frac{\varepsilon|\omega_2|}{1 + |\omega_1| + |\omega_2|}.$$

Since $k\omega_3 = k\alpha\omega_1 + k\beta\omega_2$ we find, by the triangle inequality,

$$|k\omega_3 - h_1\omega_1 - h_2\omega_2| < \frac{\varepsilon(|\omega_1| + |\omega_2|)}{1 + |\omega_1| + |\omega_2|} < \varepsilon.$$

Thus $k\omega_3 - h_1\omega_1 - h_2\omega_2$ is a nonzero period with modulus $< \varepsilon$. $\qquad\square$

Note. In Chapter 1 we showed that a function with arbitrarily small nonzero periods is constant on every open connected subset in which it is analytic. Therefore, by Theorem 7.13, the only meromorphic functions with three independent periods are constant functions.

Further applications of Kronecker's theorem are given in the next chapter.

Exercises for Chapter 7

1. Prove the following extension of Dirichlet's approximation theorem.

Given n real numbers $\theta_1, \ldots, \theta_n$ and given an integer $N \geq 1$, there exist integers $h_1, \ldots, h_n$ and k, with $1 \leq k \leq N^n$, such that

$$|k\theta_i - h_i| < \frac{1}{N} \quad \text{for } i = 1, 2, \ldots, n.$$

2. (a) Given n real numbers $\theta_1, \ldots, \theta_n$, prove that there exist integers $h_1, \ldots, h_n$ and $k > 0$ such that

$$\left|\theta_i - \frac{h_i}{k}\right| < \frac{1}{k^{1 + 1/n}} \quad \text{for } i = 1, 2, \ldots, n.$$

(b) If at least one of the θ_i is irrational, prove that there is an infinite set of n-tuples $(h_1/k, \ldots, h_n/k)$ satisfying the inequalities in (a).

3. This exercise gives another extension of Dirichlet's approximation theorem. Given m linear forms,

$$L_i = a_{i1}x_1 + \cdots + a_{in}x_n - y_i, \qquad i = 1, 2, \ldots, m,$$

in $n + m$ variables $x_1, \ldots, x_n, y_1, \ldots, y_m$, prove that for each integer $N > 1$ there exists integers $x_1, \ldots, x_n, y_1, \ldots, y_m$ such that

$$|L_i| < \frac{1}{N} \quad \text{for } i = 1, 2, \ldots, m$$

and $0 < \max\{|x_1|, \ldots, |x_n|\} \le N^{m/n}$. *Hint*: Let $M_j = a_{j1}x_1 + \cdots + a_{jn}x_n$ and examine the points $(\{M_1\}, \ldots, \{M_m\})$ in the unit cube in m-space, where $\{M_j\} = M_j - [M_j]$.

4. Let θ be irrational, $0 < \theta < 1$. Then θ lies between two consecutive Farey fractions, say

$$\frac{a}{b} < \theta < \frac{c}{d}.$$

(a) Prove that either $\theta - a/b < 1/(2b^2)$ or $c/d - \theta < 1/(2d^2)$.
(b) Deduce that there exist infinitely many fractions h/k with $(h, k) = 1$ and $k > 0$ such that

$$\left| \theta - \frac{h}{k} \right| < \frac{1}{2k^2}.$$

5. Let $\alpha = (1 + \sqrt{5})/2$. This exercise shows that the inequality

(20)
$$\left| \alpha - \frac{h}{k} \right| < \frac{c}{k^2}$$

has only a finite number of solutions in integers h and k with $k > 0$ if $0 < c < 1/\sqrt{5}$.

(a) Let $\beta = \alpha - \sqrt{5}$ so that α and β are roots of the equation $x^2 - x - 1 = 0$. Show that for any integers h and k with $k > 0$ we have

$$\frac{1}{k^2} \le \left| \alpha - \frac{h}{k} \right| \left| \beta - \frac{h}{k} \right|$$

and deduce that

$$\frac{1}{k^2} \le \left| \alpha - \frac{h}{k} \right| \left(\left| \alpha - \frac{h}{k} \right| + \sqrt{5} \right).$$

(b) If (20) has infinitely many solutions h/k with $k > 0$, say $h_1/k_1, h_2/k \cdot, \ldots$, show that $k_n \to \infty$ as $n \to \infty$ and use part (a) to prove that $c \ge 1/\sqrt{5}$.

6. In Lemma 2, define

$$L = \limsup_{t \to +\infty} |F(t)| \quad \text{instead of } L = \sup_{-\infty < t < \infty} |F(t)|.$$

Prove that the equation $L = n + 1$ is equivalent to the following statement: For every $\varepsilon > 0$ and every $T > 0$ there exists a real $t > T$ and integers $h_1, \ldots, h_n$ such that $|t\theta_i - h_i - \alpha_i| < \varepsilon$ for every $i = 1, 2, \ldots, n$.

7. Prove that the multiplier t in the first form of Kronecker's theorem can be taken positive and arbitrarily large. That is, under the hypotheses of Theorem 7.9, if $T > 0$ is given there exists a real $t > T$ satisfying the n inequalities $|t\theta_i - h_i - \alpha_i| < \varepsilon$. Show also that the integer multiplier k in the second form of Kronecker's theorem can be taken positive and arbitrarily large.

General Dirichlet series and Bohr's equivalence theorem

8

8.1 Introduction

This chapter treats a class of series, called general Dirichlet series, which includes both power series and ordinary Dirichlet series as special cases. Most of the chapter is devoted to a method developed by Harald Bohr [6] in 1919 for studying the set of values taken by Dirichlet series in a half-plane. Bohr introduced an equivalence relation among Dirichlet series and showed that equivalent Dirichlet series take the same set of values in certain half-planes. The theory uses Kronecker's approximation theorem discussed in the previous chapter. At the end of the chapter applications are given to the Riemann zeta function and to Dirichlet L-functions.

8.2 The half-plane of convergence of general Dirichlet series

Definition. Let $\{\lambda(n)\}$ be a strictly increasing sequence of real numbers such that $\lambda(n) \to +\infty$ as $n \to \infty$. A series of the form

$$\sum_{n=1}^{\infty} a(n)e^{-s\lambda(n)}$$

is called a *general Dirichlet series*. The numbers $\lambda(n)$ are called the *exponents* of the series, and the numbers $a(n)$ are called its *coefficients*.

As usual, we write $s = \sigma + it$ where σ and t are real.

Note. When $\lambda(n) = \log n$ then $e^{-s\lambda(n)} = n^{-s}$ and we obtain the ordinary Dirichlet series $\sum a(n)n^{-s}$. When $\lambda(n) = n$ the series becomes a power series in x, where $x = e^{-s}$.

A general Dirichlet series is analogous to the Laplace transform of a function, $\int_0^\infty f(t)e^{-st}\,dt$. As a matter of fact, both Dirichlet series and Laplace transforms are special cases of the *Laplace–Stieltjes transform*, $\int_0^\infty e^{-st}\,d\alpha(t)$. When $\alpha(t)$ has a continuous derivative $\alpha'(t) = f(t)$ this gives the Laplace transform of f. When α is a step function with jump $a(n)$ at the point $\lambda(n)$ the integral becomes the general Dirichlet series $\sum a(n)e^{-s\lambda(n)}$. Much of what we do here can be extended to Laplace–Stieltjes transforms, but we shall not deal with these generalizations.

As is the case with ordinary Dirichlet series, each general Dirichlet series has associated with it an abscissa σ_c of convergence and an abscissa σ_a of absolute convergence. We could argue as in Chapter 11 of [4] to prove the existence of σ_c and σ_a. Instead we give a different method of proof which also expresses σ_c and σ_a in terms of the exponents $\lambda(n)$ and the coefficients $a(n)$.

Theorem 8.1. *Assume that the series $\sum a(n)e^{-s\lambda(n)}$ converges for some s with positive real part, say for $s = s_0$ with $\sigma_0 > 0$. Let*

$$L = \limsup_{n \to \infty} \frac{\log|\sum_{k=1}^n a(k)|}{\lambda(n)}.$$

Then $L \leq \sigma_0$. Moreover, the series converges in the half-plane $\sigma > L$, and the convergence is uniform on every compact subset of the half-plane $\sigma > L$.

PROOF. First we prove that $L \leq \sigma_0$. Let $A(n)$ denote the partial sums of the coefficients,

$$A(n) = \sum_{k=1}^n a(k).$$

Note that $\lambda(n) > 0$ for all sufficiently large n. If we prove that for every $\varepsilon > 0$ we have

(1) $$\log|A(n)| < (\sigma_0 + \varepsilon)\lambda(n)$$

for all sufficiently large n, then it follows that

$$\frac{\log|A(n)|}{\lambda(n)} < \sigma_0 + \varepsilon$$

for these n, so $L \leq \sigma_0 + \varepsilon$, hence $L \leq \sigma_0$. Now relation (1) is equivalent to the inequality

(2) $$|A(n)| < e^{(\sigma_0 + \varepsilon)\lambda(n)}.$$

To prove (2) we introduce the partial sums

$$S(n) = \sum_{k=1}^n a(k)e^{-s_0\lambda(k)}.$$

162

The $S(n)$ are bounded since the series $\sum_{k=1}^{\infty} a(k)e^{-s_0\lambda(k)}$ converges. Suppose that $|S(n)| < M$ for all n. To express $A(n)$ in terms of the $S(n)$ we use partial summation:

$$A(n) = \sum_{k=1}^{n} a(k) = \sum_{k=1}^{n} a(k)e^{-s_0\lambda(k)}e^{s_0\lambda(k)}$$

$$= \sum_{k=1}^{n} \{S(k) - S(k-1)\}e^{s_0\lambda(k)},$$

provided $S(0) = 0$. Thus

$$A(n) = \sum_{k=1}^{n} S(k)e^{s_0\lambda(k)} - \sum_{k=1}^{n-1} S(k)e^{s_0\lambda(k+1)}$$

$$= \sum_{k=1}^{n-1} S(k)\{e^{s_0\lambda(k)} - e^{s_0\lambda(k+1)}\} + S(n)e^{s_0\lambda(n)}.$$

Hence

$$|A(n)| < M \sum_{k=1}^{n-1} |e^{s_0\lambda(k)} - e^{s_0\lambda(k+1)}| + Me^{\sigma_0\lambda(n)}.$$

But

$$\sum_{k=1}^{n-1} |e^{s_0\lambda(k)} - e^{s_0\lambda(k+1)}| = \sum_{k=1}^{n-1} \left| s_0 \int_{\lambda(k)}^{\lambda(k+1)} e^{s_0 u}\, du \right| \leq |s_0| \sum_{k=1}^{n-1} \int_{\lambda(k)}^{\lambda(k+1)} e^{\sigma_0 u}\, du$$

$$= |s_0| \int_{\lambda(1)}^{\lambda(n)} e^{\sigma_0 u}\, du = \frac{|s_0|}{\sigma_0}(e^{\sigma_0\lambda(n)} - e^{\sigma_0\lambda(1)}) < \frac{|s_0|}{\sigma_0}e^{\sigma_0\lambda(n)}.$$

Thus

$$|A(n)| < M\left(1 + \frac{|s_0|}{\sigma_0}\right)e^{\sigma_0\lambda(n)}.$$

Now $\lambda(n) \to \infty$ as $n \to \infty$ so

$$e^{\varepsilon\lambda(n)} > M\left(1 + \frac{|s_0|}{\sigma_0}\right)$$

if n is sufficiently large. Hence for these n we have $|A(n)| < e^{(\sigma_0 + \varepsilon)\lambda(n)}$, which proves (2) and hence (1). This proves that $L \leq \sigma_0$.

Now we prove that the series converges for all s with $\sigma > L$. Consider any section of the series $\sum a(n)e^{-s\lambda(n)}$, say $\sum_{n=a}^{b}$. We shall use the Cauchy convergence criterion to show that this section can be made small when a and b are sufficiently large. We estimate the size of such a section by using partial summation to compare it to the partial sums $A(n) = \sum_{k=1}^{n} a(k)$. We have

$$\sum_{n=a}^{b} a(n)e^{-s\lambda(n)} = \sum_{n=a}^{b} \{A(n) - A(n-1)\}e^{-s\lambda(n)}$$

$$= \sum_{n=a}^{b} A(n)\{e^{-s\lambda(n)} - e^{-s\lambda(n+1)}\} + A(b)e^{-s\lambda(b+1)}$$

$$- A(a-1)e^{-s\lambda(a)}.$$

This relation holds for any choice of s, a and b. Now suppose s is any complex number with $\sigma > L$. Let $\varepsilon = \frac{1}{2}(\sigma - L)$. Then $\varepsilon > 0$ and $\sigma = L + 2\varepsilon$. By the definition of L, for this ε there is an integer $N(\varepsilon)$ such that for all $n \geq N(\varepsilon)$ we have

$$\frac{\log|A(n)|}{\lambda(n)} < L + \varepsilon.$$

We can also assume that $\lambda(n) > 0$ for $n \geq N(\varepsilon)$. Hence

$$|A(n)| < e^{(L+\varepsilon)\lambda(n)} \quad \text{for all } n \geq N(\varepsilon).$$

If we choose $b \geq a > N(\varepsilon)$ we get the estimate

$$\left| \sum_{n=a}^{b} a(n)e^{-s\lambda(n)} \right| \leq \sum_{n=a}^{b} e^{(L+\varepsilon)\lambda(n)} |e^{-s\lambda(n)} - e^{-s\lambda(n+1)}|$$

$$+ e^{(L+\varepsilon)\lambda(b+1)}e^{-\sigma\lambda(b+1)} + e^{(L+\varepsilon)\lambda(a)}e^{-\sigma\lambda(a)}.$$

The last two terms are $e^{-\varepsilon\lambda(b+1)} + e^{-\varepsilon\lambda(a)}$ since $L + \varepsilon - \sigma = -\varepsilon$. Now we estimate the sum by writing

$$|e^{-s\lambda(n)} - e^{-s\lambda(n+1)}| = \left| -s \int_{\lambda(n)}^{\lambda(n+1)} e^{-su} \, du \right| \leq |s| \int_{\lambda(n)}^{\lambda(n+1)} e^{-\sigma u} \, du$$

so

$$\sum_{n=a}^{b} e^{(L+\varepsilon)\lambda(n)} |e^{-s\lambda(n)} - e^{-s\lambda(n+1)}| \leq |s| \sum_{n=a}^{b} e^{(L+\varepsilon)\lambda(n)} \int_{\lambda(n)}^{\lambda(n+1)} e^{-\sigma u} \, du$$

$$\leq |s| \sum_{n=a}^{b} \int_{\lambda(n)}^{\lambda(n+1)} e^{-\sigma u} e^{(L+\varepsilon)u} \, du = |s| \sum_{n=a}^{b} \int_{\lambda(n)}^{\lambda(n+1)} e^{-\varepsilon u} \, du$$

$$= |s| \int_{\lambda(a)}^{\lambda(b+1)} e^{-\varepsilon u} \, du = \frac{|s|}{\varepsilon} (e^{-\varepsilon\lambda(a)} - e^{-\varepsilon\lambda(b+1)}).$$

Thus we have

$$\left| \sum_{n=a}^{b} a(n)e^{-s\lambda(n)} \right| \leq \frac{|s|}{\varepsilon} (e^{-\varepsilon\lambda(a)} - e^{-\varepsilon\lambda(b+1)}) + e^{-\varepsilon\lambda(b+1)} + e^{-\varepsilon\lambda(a)}.$$

Each term on the right tends to 0 as $a \to \infty$, so the Cauchy criterion shows that the series converges for all s with $\sigma > L$. This completes the proof. Note also that this proves *uniform* convergence on any compact subset of the half-plane $\sigma > L$. $\qquad\qquad\qquad\qquad\qquad\qquad\qquad\qquad\square$

Theorem 8.2. *Assume the series $\sum a(n)e^{-s\lambda(n)}$ converges for some s with $\sigma > 0$ but diverges for all s with $\sigma < 0$. Then the number*

$$L = \limsup_{n \to \infty} \frac{\log|\sum_{k=1}^{n} a(k)|}{\lambda(n)}$$

is the abscissa of convergence of the series. In other words, the series converges for all s with $\sigma > L$ and diverges for all s with $\sigma < L$.

PROOF. We know from Theorem 8.1 that the series converges for all s with $\sigma > L$ and that L cannot be negative. Let S be the set of all $\sigma > 0$ such that the series converges for some s with real part σ. The set S is nonempty and bounded below. Let σ_c be the greatest lower bound of S. Then $\sigma_c > 0$. Each σ in S satisfies $L \leq \sigma$ hence $L \leq \sigma_c$. If we had $\sigma_c > L$ there would be a σ in the interval $L < \sigma < \sigma_c$. For this σ we would also have convergence for all s with real part σ (by Theorem 8.1) contradicting the definition of σ_c. Hence $\sigma_c = L$. But the definition of σ_c shows that the series diverges for all s with $0 \leq \sigma < L$. By hypothesis it also diverges for all s with $\sigma < 0$. Hence it diverges for all s with $\sigma < L$. This completes the proof. $\qquad\square$

As a corollary we have:

Theorem 8.3. *Assume the series $\sum a(n)e^{-s\lambda(n)}$ converges absolutely for some s with $\sigma > 0$ but diverges for all s with $\sigma < 0$. Then the number*

$$\sigma_a = \limsup_{n \to \infty} \frac{\log \sum_{k=1}^{n} |a(k)|}{\lambda(n)}$$

is the abscissa of absolute convergence of the series.

PROOF. Let A be the abscissa of convergence of the series $\sum |a(n)|e^{-s\lambda(n)}$. Then, by Theorem 8.2,

$$A = \limsup_{n \to \infty} \frac{\log \sum_{k=1}^{n} |a(k)|}{\lambda(n)}.$$

We wish to prove that $\sum |a(n)|e^{-\sigma\lambda(n)}$ converges if $\sigma > A$ and diverges if $\sigma < A$. Clearly if $\sigma > A$ then the point $s = \sigma$ is within the half-plane of convergence of $\sum |a(n)|e^{-s\lambda(n)}$ so $\sum |a(n)|e^{-\sigma\lambda(n)}$ converges.

Now suppose $\sum |a(n)|e^{-\sigma\lambda(n)}$ converges for some $\sigma < A$. Then the series $\sum |a(n)|e^{-s\lambda(n)}$ converges absolutely for each s with real part σ so, in particular it *converges* for all these s, contradicting the fact that A is the abscissa of convergence of $\sum |a(n)|e^{-s\lambda(n)}$. $\qquad\qquad\qquad\square$

8.3 Bases for the sequence of exponents of a Dirichlet series

The rest of this chapter is devoted to a detailed study of Harald Bohr's theory with applications to the Riemann zeta-function and Dirichlet's L-series. The first notion we need is that of a *basis* for the sequence of exponents of a Dirichlet series.

Definition. Let $\Lambda = \{\lambda(n)\}$ be an infinite sequence of distinct real numbers. By a basis of the set Λ we shall mean a finite or countably infinite sequence $B = \{\beta(n)\}$ of real numbers satisfying the following three conditions:

(a) The sequence B is linearly independent over the rationals. That is, for all $m \geq 1$, if

$$\sum_{k=1}^{m} r_k \beta(k) = 0$$

with rational multipliers r_k, then each $r_k = 0$.

(b) Each $\lambda(n)$ is expressible as a finite linear combination of terms of B, say

$$\lambda(n) = \sum_{k=1}^{q(n)} r_{n,k} \beta(k)$$

where the $r_{n,k}$ are rational and the number of summands $q(n)$ depends on n. (By condition (a), if $\lambda(n) \neq 0$ this representation is unique.)

(c) Each $\beta(n)$ is expressible as a finite linear combination of terms of Λ, say

$$\beta(n) = \sum_{k=1}^{m(n)} t_{n,k} \lambda(k)$$

where the $t_{n,k}$ are rational and $m(n)$ depends on n.

EXAMPLE 1. Let Λ be the set of all rational numbers. Then $B = \{1\}$ is a basis.

EXAMPLE 2. Let $\Lambda = \{\log n\}$. Then $B = \{\log p_n\}$ is a basis, where p_n is the nth prime. It is easy to verify properties (a), (b) and (c). For independence we note that

$$\sum_{k=1}^{q} r_k \log p_k = 0 \quad \text{implies} \quad p_1^{r_1} \cdots p_q^{r_q} = 1 \quad \text{so} \quad r_1 = \cdots = r_q = 0.$$

To express each $\lambda(n)$ in terms of the basis elements we factor n and compute $\log n$ as a linear combination of the logarithms of its prime factors. Property (c) is trivially satisfied since B is a subsequence of Λ.

Theorem 8.4. *Every sequence Λ has a subsequence which is a basis for Λ.*

PROOF. Construct a basis as follows. For the first basis element take $\lambda(n_1)$, the first nonzero λ (either $\lambda(1)$ or $\lambda(2)$), and call this $\beta(1)$. Now delete the remaining elements of Λ that are rational multiples of $\beta(1)$. If this exhausts all of Λ take $B = \{\beta(1)\}$. If not, let $\lambda(n_2)$ denote the first remaining λ, take $\beta(2) = \lambda(n_2)$, and strike out the remaining elements of Λ which are rational linear combinations of $\beta(1)$ and $\beta(2)$. Continue in this fashion to obtain a sequence $B = (\beta(1), \beta(2), \ldots) = (\lambda(n_1), \lambda(n_2), \ldots)$. It is easy to verify that B is a basis for Λ. Property (a) holds by construction, since each β was chosen to be independent of the earlier elements. To verify (b) we note that every λ is either an element of B or a rational linear combination of a finite number of elements of B. Finally, (c) holds trivially since B is a subsequence of Λ. $\quad\square$

Note. Every sequence Λ has infinitely many bases.

8.4 Bohr matrices

It is convenient to express these concepts in matrix notation. We display the sequences Λ and B as column matrices, using an infinite column matrix for Λ and a finite or infinite column matrix for B, according as B is a finite or infinite sequence.

We also consider finite or infinite square matrices $R = (r_{ij})$ with rational entries. If R is infinite we require that all but a finite number of entries in each row be zero. Such rational square matrices will be called *Bohr matrices*.

We define matrix addition and multiplication of two infinite Bohr matrices as for finite matrices. Note that a sum or product of two Bohr matrices is another Bohr matrix. Also, the product RB of a Bohr matrix R with an infinite column matrix B is another infinite column matrix Γ. Moreover, we have the associative property $(R_1 R_2)B = R_1(R_2 B)$ if R_1 and R_2 are Bohr matrices and B is an infinite column matrix.

In matrix notation, the definition of basis takes the following form. B is called a basis for Λ if it satisfies the following three conditions:

(a) If $RB = 0$ for some Bohr matrix R, then $R = 0$.
(b) There exists a Bohr matrix R such that $\Lambda = RB$.
(c) There exists a Bohr matrix T such that $B = T\Lambda$.

The relation between two bases B and Γ of the same sequence Λ can be expressed as follows:

Theorem 8.5. *If Λ has two bases B and Γ, then there exists a Bohr matrix A such that $\Gamma = AB$.*

PROOF. There exist Bohr matrices R and T such that $\Gamma = T\Lambda$ and $\Lambda = RB$. Hence $\Gamma = T(RB) = (TR)B = AB$ where $A = TR$. $\quad\square$

Theorem 8.6. *Let B and Γ be two bases for Λ, and write $\Gamma = AB$, $\Lambda = R_B B$, $\Lambda = R_\Gamma \Gamma$, where A, R_B, R_Γ are Bohr matrices. Then $R_B = R_\Gamma A$.*

Note. If we write Λ/B for R_B, Λ/Γ for R_Γ and Γ/B for A, this last equation states that

$$\frac{\Lambda}{B} = \frac{\Lambda}{\Gamma} \cdot \frac{\Gamma}{B}.$$

PROOF. We have $\Lambda = R_B B$ and $\Lambda = R_\Gamma \Gamma = R_\Gamma AB$. Hence $R_B B = R_\Gamma AB$, so $(R_B - R_\Gamma A)B = 0$. Since $R_B - R_\Gamma A$ is a Bohr matrix and B is a basis, we must have $R_B - R_\Gamma A = 0$. $\qquad\square$

8.5 The Bohr function associated with a Dirichlet series

To every Dirichlet series $f(s) = \sum_{n=1}^{\infty} a(n)e^{-s\lambda(n)}$ we associate a function $F(z_1, z_2, \ldots)$ of countably many complex variables $z_1, z_2, \ldots$ as follows. Let Z denote the column matrix with entries $z_1, z_2, \ldots$. Let $B = \{\beta(n)\}$ be a basis for the sequence $\Lambda = \{\lambda(n)\}$ of exponents, and write $\Lambda = RB$, where R is a Bohr matrix.

Definition. The Bohr function $F(Z) = F(z_1, z_2, \ldots)$ associated with $f(s)$, relative to the basis B, is the series

$$F(Z) = \sum_{n=1}^{\infty} a(n)e^{-(RZ)_n},$$

where $(RZ)_n$ denotes the nth entry of the column matrix RZ.

In other words, if

$$\lambda(n) = \sum_{k=1}^{q(n)} r_{n,k}\beta(k)$$

then

$$F(z_1, z_2, \ldots) = \sum_{n=1}^{\infty} a(n)e^{-(r_{n,1}z_1 + \cdots + r_{n,q(n)}z_{q(n)})}.$$

Note that the formal substitution $z_m = s\beta_m$ gives $Z = sB$, $RZ = sRB = s\Lambda$, so $(RZ)_n = s\lambda(n)$ and hence

$$F(sB) = \sum_{n=1}^{\infty} a(n)e^{-s\lambda(n)} = f(s).$$

In other words, the Dirichlet series $f(s)$ arises from $F(Z)$ by a special choice of the variables $z_1, z_2, \ldots$. Therefore, if the Dirichlet series $f(s)$ converges for $s = \sigma + it$ the associated Bohr series $F(Z)$ also converges when $Z = sB$.

Moreover, if the Dirichlet series $f(s)$ converges absolutely for $s = \sigma + it$ then the Bohr series $F(Z)$ converges absolutely for any choice of $z_1, z_2, \ldots$ with Re $z_n = \sigma\beta(n)$ for all n. To see this we note that if Re $z_n = \sigma\beta(n)$ then Re $Z = \sigma B$ so

$$\sum_{n=1}^{\infty} |a(n)e^{-(RZ)_n}| = \sum_{n=1}^{\infty} |a(n)|e^{-\sigma(RB)_n} = \sum_{n=1}^{\infty} |a(n)|e^{-\sigma\lambda(n)}.$$

To emphasize the dependence of the Bohr function on the basis B we sometimes write $\Lambda = R_B B$ and

$$F_B(Z) = \sum_{n=1}^{\infty} a(n)e^{-(R_B Z)_n}.$$

Bohr functions F_B and F_Γ corresponding to different bases are related by the following theorem.

Theorem 8.7. *Let B and Γ be two bases for Λ and write $\Gamma = AB$ for some Bohr matrix A. Then*

$$F_B(Z) = F_\Gamma(AZ).$$

PROOF. By Theorem 8.6 we have

$$\Lambda = R_B B = R_\Gamma \Gamma, \quad \text{where } R_B = R_\Gamma A.$$

Hence

$$F_B(Z) = \sum_{n=1}^{\infty} a(n)e^{-(R_B Z)_n} = \sum_{n=1}^{\infty} a(n)\exp\{-(R_\Gamma AZ)_n\} = F_\Gamma(AZ). \qquad \square$$

Definition. Assume the Dirichlet series $f(s) = \sum_{n=1}^{\infty} a(n)e^{-s\lambda(n)}$ converges absolutely for some $s = \sigma + it$. We define $U_f(\sigma; B)$ to be the set of values taken on by the associated Bohr function, relative to the basis B, when Re $Z = \sigma B$. Thus,

$$U_f(\sigma; B) = \{F(Z) : \text{Re } Z = \sigma B\}.$$

The next theorem shows that this set is independent of the basis B.

Theorem 8.8. *If B and Γ are two bases for Λ then $U_f(\sigma; B) = U_f(\sigma; \Gamma)$.*

PROOF. Choose any value $F_B(Z)$ in $U_f(\sigma; B)$, so that Re $Z = \sigma B$. By Theorem 8.7 we have $F_B(Z) = F_\Gamma(AZ)$, where $\Gamma = AB$. But

$$\text{Re } AZ = A \text{ Re } Z = A\sigma B = \sigma AB = \sigma\Gamma$$

so $F_B(Z) \in U_f(\sigma; \Gamma)$. This proves $U_f(\sigma; B) \subseteq U_f(\sigma; \Gamma)$, and a similar argument gives $U_f(\sigma; \Gamma) \subseteq U_f(\sigma; B)$. $\qquad \square$

Note. Since $U_f(\sigma; B)$ is independent of the basis B we designate the set $U_f(\sigma; B)$ simply by $U_f(\sigma)$.

8.6 The set of values taken by a Dirichlet series $f(s)$ on a line $\sigma = \sigma_0$

This section relates the set $U_f(\sigma_0)$ with the set of values taken by the Dirichlet series $f(s)$ on the line $\sigma = \sigma_0$.

Definition. If the Dirichlet series $f(s) = \sum_{n=1}^{\infty} a(n)e^{-s\lambda(n)}$ converges absolutely for $\sigma = \sigma_0$ we let

$$V_f(\sigma_0) = \{f(\sigma_0 + it): -\infty < t < +\infty\}$$

denote the set of values taken by $f(s)$ on the line $\sigma = \sigma_0$.

Since $f(s)$ can be obtained from its Bohr function $F(Z)$ by putting $Z = \sigma B$, it follows that $V_f(\sigma_0) \subseteq U_f(\sigma_0)$. Now we prove an inclusion relation in the other direction.

Theorem 8.9. *Assume $\sigma_0 > \sigma_a$, where σ_a is the abscissa of absolute convergence of a Dirichlet series $f(s)$. Then the closure of $V_f(\sigma_0)$ contains $U_f(\sigma_0)$. That is, we have*

$$V_f(\sigma_0) \subseteq U_f(\sigma_0) \subseteq \overline{V_f(\sigma_0)}, \quad \text{and hence } \overline{U_f(\sigma_0)} = \overline{V_f(\sigma_0)}.$$

PROOF. The closure $\overline{V_f(\sigma_0)}$ is the set of adherent points of $V_f(\sigma_0)$. We are to prove that every point u in $U_f(\sigma_0)$ is an adherent point of $V_f(\sigma_0)$. In other words, given u in $U_f(\sigma_0)$ and given $\varepsilon > 0$ we will prove that there exists a v in $V_f(\sigma_0)$ such that $|u - v| < \varepsilon$. Since $v = f(\sigma_0 + it)$ for some t, we are to prove that there exists a real t such that

$$|f(\sigma_0 + it) - u| < \varepsilon.$$

Since $u \in U_f(\sigma_0)$ we have $u = F(z_1, z_2, \ldots)$ where $z_n = \sigma_0 \beta(n) + iy_n$. Hence

$$Z = \sigma_0 B + iY, \qquad RZ = \sigma_0 RB + iRY = \sigma_0 \Lambda + iRY,$$

so

$$(RZ)_n = \sigma_0 \lambda(n) + i(RY)_n = \sigma_0 \lambda(n) + i\mu_n,$$

say. Therefore

$$u = \sum_{n=1}^{\infty} a(n)e^{-\sigma_0\lambda(n)}e^{-i\mu_n}.$$

On the other hand, we have

$$f(\sigma_0 + it) = \sum_{n=1}^{\infty} a(n)e^{-\sigma_0\lambda(n)}e^{-it\lambda(n)},$$

hence

$$f(\sigma_0 + it) - u = \sum_{n=1}^{\infty} a(n)e^{-\sigma_0\lambda(n)}(e^{-it\lambda(n)} - e^{-i\mu_n}).$$

The idea of the proof from here on is as follows: First we split the sum into two parts, $\sum_{n=1}^{N} + \sum_{n=N+1}^{\infty}$. We choose N so the second part $\sum_{n=N+1}^{\infty}$ is small, say its absolute value is $< \frac{1}{2}\varepsilon$. This is possible by absolute convergence. Then we show that the first part can be made small by choosing t properly. The idea is to choose t to make every exponential $e^{-it\lambda(n)}$ very close to $e^{-i\mu_n}$ simultaneously for every $n = 1, 2, \ldots, N$. Then each factor $e^{-it\lambda(n)} - e^{-i\mu_n}$ will be small, and since there are only N terms, the whole sum will be small.

Now we discuss the details. For the given ε, choose N so that

$$\left| \sum_{n=N+1}^{\infty} a(n)e^{-\sigma_0\lambda(n)}(e^{-it\lambda(n)} - e^{-i\mu_n}) \right| < \frac{\varepsilon}{2}.$$

Then we have

$$|f(\sigma_0 + it) - u| < \left| \sum_{n=1}^{N} a(n)e^{-\sigma_0\lambda(n)}(e^{-it\lambda(n)} - e^{-i\mu_n}) \right| + \frac{\varepsilon}{2}.$$

This holds for any choice of t. We wish to choose t to make the first sum $< \frac{1}{2}\varepsilon$. Since $|e^{it\lambda(n)}| = 1$ we can rewrite the sum in question as follows:

$$\left| \sum_{n=1}^{N} a(n)e^{-\sigma_0\lambda(n)}(e^{-it\lambda(n)} - e^{-i\mu_n}) \right| = \left| \sum_{n=1}^{N} e^{-it\lambda(n)}a(n)e^{-\sigma_0\lambda(n)}(1 - e^{i(t\lambda(n)-\mu_n)}) \right|$$

$$\leq \sum_{n=1}^{N} |a(n)|e^{-\sigma_0\lambda(n)}|e^{i(t\lambda(n)-\mu_n)} - 1|.$$

Let $M = 1 + \sum_{n=1}^{N} |a(n)|e^{-\sigma_0\lambda(n)}$. For the given ε there is a $\delta > 0$ such that

(3) $$|e^{ix} - 1| < \frac{\varepsilon}{2M} \quad \text{if } |x| < \delta.$$

Suppose we could choose a real t and integers $k_1, \ldots, k_N$ such that

(4) $$t\lambda(n) - \mu_n = 2\pi k_n + x_n$$

where $|x_n| < \delta$ for $n = 1, 2, \ldots, N$. Then for this t we would have

$$e^{i(t\lambda(n) - \mu_n)} = e^{2\pi i k_n + ix_n} = e^{ix_n}.$$

By (3), this would give us

$$|e^{i(t\lambda(n) - \mu_n)} - 1| < \frac{\varepsilon}{2M},$$

and hence

$$\sum_{n=1}^{N} |a(n)| e^{-\sigma_0 \lambda(n)} |e^{i(t\lambda(n) - \mu_n)} - 1| < \frac{\varepsilon}{2M} \sum_{n=1}^{N} |a(n)| e^{-\sigma_0 \lambda(n)} < \frac{\varepsilon}{2}.$$

Thus, the proof will be complete if we can find t and integers $k_1, \ldots, k_N$ to satisfy (4). If the $\lambda(n)$ were linearly independent over the integers we could apply Kronecker's theorem to $\lambda(1), \ldots \lambda(N)$ and obtain (4). However the $\lambda(n)$ are not necessarily independent so instead we apply Kronecker's theorem to the following system:

$$\theta_1, \theta_2, \ldots, \theta_Q, \qquad \alpha_1, \alpha_2, \ldots, \alpha_Q,$$

where

$$\theta_n = \frac{\beta(n)}{2\pi D}, \qquad \alpha_n = \frac{y_n}{2\pi D}.$$

The $\beta(n)$ are the elements of the basis B used to define $F(Z)$, and the y_n are the imaginary parts of the numbers z_n which determine u. The integers Q and D are determined as follows. We express Λ in terms of B by writing

$$\lambda(n) = r_{n,1}\beta(1) + \cdots + r_{n, q(n)}\beta(q(n)).$$

Then Q is the largest of the integers $q(1), \ldots, q(N)$, and D is the least common multiple of the denominators of the rational numbers $r_{i,j}$ that arise from the $\lambda(n)$ appearing in the sum. There are at most $q(1) + \cdots + q(N)$ such numbers $r_{i,j}$. The numbers θ_n are linearly independent over the integers because B is a basis.

By Kronecker's theorem a real t and integers $h_1, \ldots, h_Q$ exist such that

$$|t\theta_k - \alpha_k - h_k| < \frac{\delta}{2\pi D A},$$

where

$$A = \sum_{n=1}^{N} \sum_{j=1}^{q(n)} |r_{n,j}|.$$

For this t we have $|2\pi Dt\theta_k - 2\pi D\alpha_k - 2\pi Dh_k| < \delta/A$, or

$$|t\beta(k) - y_k - 2\pi Dh_k| < \frac{\delta}{A}.$$

Therefore $t\beta(k) - y_k = 2\pi Dh_k + \delta_k$, where $|\delta_k| < \delta/A$. Now we can write

$$t\lambda(n) - \mu_n = t\sum_{j=1}^{q(n)} r_{n,j}\beta(j) - \sum_{j=1}^{q(n)} r_{n,j}y_j$$

$$= \sum_{j=1}^{q(n)} r_{n,j}(t\beta(j) - y_j) = \sum_{j=1}^{q(n)} r_{n,j}(2\pi Dh_j + \delta_j)$$

$$= 2\pi \sum_{j=1}^{q(n)} h_j Dr_{n,j} + \sum_{j=1}^{q(n)} \delta_j r_{n,j}$$

$$= 2\pi k_n + x_n$$

where k_n is an integer and $|x_n| < (\delta/A) \sum_{j=1}^{q(n)} |r_{n,j}| < \delta$. But this means we have found a real t and integers $k_1, \ldots, k_N$ to satisfy (4), so the proof is complete. $\qquad\qquad\square$

8.7 Equivalence of general Dirichlet series

Consider two general Dirichlet series with the same sequence of exponents Λ, say

$$\sum_{n=1}^{\infty} a(n)e^{-s\lambda(n)} \quad \text{and} \quad \sum_{n=1}^{\infty} b(n)e^{-s\lambda(n)}.$$

Let $B = \{\beta(n)\}$ be a basis for Λ and write $\Lambda = RB$, where R is a Bohr matrix.

Definition. We say the two series are equivalent, relative to the basis B, and we write

$$\sum_{n=1}^{\infty} a(n)e^{-s\lambda(n)} \sim \sum_{n=1}^{\infty} b(n)e^{-s\lambda(n)}$$

if there exists a finite or infinite sequence of real numbers $Y = \{y_n\}$ such that

$$b(n) = a(n)e^{ix_n}$$

where $X = \{x_n\} = RY.$

In other words, if we write

$$\lambda(n) = \sum_{k=1}^{q(n)} r_{n,k}\beta(k),$$

equivalence means that for some sequence $\{y_n\}$ we have

$$b(n) = a(n) \exp\left(i \sum_{k=1}^{q(n)} r_{n,k}y_k\right).$$

Theorem 8.10. *Two equivalent Dirichlet series have the same abscissa of absolute convergence. Moreover, the relation $\sim$ just defined is independent of the basis B.*

PROOF. Equivalence implies $|b(n)| = |a(n)|$ so the series have the same abscissa of absolute convergence.

Now let B and Γ be two bases for Λ, and assume that two series are equivalent with respect to B. We will show that they are also equivalent with respect to Γ.

Write $\Lambda = R_B B$. Then there is a sequence $Y = \{y_n\}$ such that $b(n) = a(n)e^{ix_n}$, where $X = \{x_n\} = R_B Y$. Now write $\Lambda = R_\Gamma \Gamma$. If we show that for some sequence $V = \{v_n\}$ we have $X = R_\Gamma V$ then the two series will be equivalent relative to Γ. The sequence

$$V = AY$$

173

has this property, where A is the Bohr matrix such that $\Gamma = AB$. In fact, we have $R_\Gamma V = R_\Gamma A Y = R_B Y = X$, since $R_\Gamma A = R_B$. This completes the proof. $\qquad\qquad\qquad\qquad\qquad\qquad\qquad\qquad\qquad\qquad\qquad\qquad\qquad$ □

Theorem 8.11. *The relation* $\sim$ *defined in the foregoing definition is an equivalence relation. That is, it is reflective, symmetric, and transitive.*

PROOF. Every series is equivalent to itself since we may take each $y_n = 0$. The corresponding x_n will then be zero.

If $b(n) = a(n)e^{ix_n}$ then $a(n) = b(n)e^{-ix_n}$. Since $X = R_B Y$ we have $-X = R_B(-Y)$ so the relation is symmetric.

To prove transitivity we may use the same basis throughout and assume that $b(n) = a(n)e^{ix_n}$, where $X = R_B Y$ for some Y, and that $a(n) = c(n)e^{iu_n}$, where $U = R_B V$ for some V. Then $b(n) = c(n)e^{i(x_n + u_n)}$ where

$$X + U = R_B Y + R_B V = R_B(Y + V).$$

This completes the proof. $\qquad\qquad\qquad\qquad\qquad\qquad\qquad\qquad\qquad\qquad\qquad\qquad$ □

8.8 Equivalence of ordinary Dirichlet series

Theorem 8.12. *Two ordinary Dirichlet series*

$$\sum_{n=1}^{\infty} \frac{a(n)}{n^s} \quad and \quad \sum_{n=1}^{\infty} \frac{b(n)}{n^s}$$

are equivalent if, and only if, there exists a completely multiplicative function f *such that*

(a) $b(n) = a(n)f(n)$ *for all* $n \geq 1$, *and*
(b) $|f(p)| = 1$ *whenever* $a(n) \neq 0$ *and* p *is a prime divisor of* n.

PROOF. For ordinary Dirichlet series the sequence of exponents $\Lambda = \{\lambda(n)\}$ is $\{\log n\}$ and for a basis we may use the sequence $B = \{\log p_n\}$, where p_n denotes the nth prime. In fact, if we use the prime-power decomposition

$$(5) \qquad\qquad\qquad\qquad n = \prod_{k=1}^{\infty} p_k^{a_{n,k}}$$

where each exponent $a_{n,k} \geq 0$, we have

$$\log n = \sum_{k=1}^{\infty} a_{n,k} \log p_k,$$

so the integer powers may be used as entries in the Bohr matrix R_B for which $\Lambda = R_B B$. In the sum and product only a finite number of the $a_{n,k}$ are nonzero.

We note that, because of the fundamental theorem of arithmetic, the numbers $a_{n,k}$ defined by (5) have the property

$$(6) \qquad\qquad\qquad\qquad a_{mn,k} = a_{m,k} + a_{n,k}.$$

Now let $A(s) = \sum a(n)n^{-s}$, $B(s) = \sum b(n)n^{-s}$. Suppose that $A(s) \sim B(s)$. Then there exists a real sequence $\{y_k\}$ such that

(7)
$$b(n) = a(n) \exp\left\{i \sum_{k=1}^{\infty} a_{n,k} y_k\right\}$$

where the integers $a_{n,k}$ are determined by equation (5). Define a function f by the equation

$$f(n) = \exp\left\{i \sum_{k=1}^{\infty} a_{n,k} y_k\right\}.$$

Property (6) implies that $f(mn) = f(m)f(n)$ for all m and n, so f is completely multiplicative. Equation (7) states that $b(n) = a(n)f(n)$, and the definition of f shows that $|f(n)| = 1$ for all n, so conditions (a) and (b) of the theorem are satisfied.

Now we prove the converse. Assume there exists a completely multiplicative function f satisfying conditions (a) and (b). We must show that there is a real sequence $\{y_k\}$ satisfying (7) for all n. First we consider those n for which $a(n) = 0$. Property (a) implies $b(n) = 0$, so equation (7) holds for such n since both sides are zero no matter how we choose the real numbers y_k. We shall now construct the sequence $\{y_k\}$ so that equation (7) also holds for those n for which $a(n) \neq 0$.

Assume, then, that n is such that $a(n) \neq 0$. We use the prime-power decomposition (5) and the completely multiplicative property of f to write

(8)
$$f(n) = \prod_{k=1}^{\infty} g(n, k),$$

where

$$g(n, k) = \begin{cases} f(p_k)^{a_{n,k}} & \text{if } p_k \mid n \\ 1 & \text{otherwise.} \end{cases}$$

Condition (b) implies that $|f(p_k)| = 1$ for each prime divisor p_k of n. Therefore for such primes we may write

$$f(p_k) = \exp(iy_k),$$

where $y_k = \arg f(p_k)$. The real numbers y_k have been defined for those k such that the prime p_k divides some n with $a(n) \neq 0$. For the remaining k (if any) we define $y_k = 0$. Thus, y_k is well-defined for every integer $k \geq 1$ and we have

$$g(n, k) = \exp(ia_{n,k} y_k)$$

for every $k \geq 1$. Equation (8) now becomes

$$f(n) = \exp\left\{i \sum_{k=1}^{\infty} a_{n,k} y_k\right\}.$$

175

This, together with property (a), shows that (7) holds for those n for which $a(n) \neq 0$. Thus, (7) holds for all n so $A(s) \sim B(s)$. This completes the proof of the theorem. $\qquad \square$

8.9 Equality of the sets $U_f(\sigma_0)$ and $U_g(\sigma_0)$ for equivalent Dirichlet series

Theorem 8.13. *Let $f(s)$ and $g(s)$ be equivalent general Dirichlet series, each of which converges absolutely for $\sigma = \sigma_0$. Then*

$$U_f(\sigma_0) = U_g(\sigma_0).$$

PROOF. Let $B = \{\beta(n)\}$ be a basis for the sequence Λ of exponents. If $f(s) = \sum a(n) e^{-s\lambda(n)}$ and $g(s) = \sum b(n) e^{-s\lambda(n)}$ then there is a real sequence $\{y_k\}$ such that

$$b(n) = a(n) \exp\left\{-i \sum_{k=1}^{q(n)} r_{n,k} y_k\right\}.$$

The Bohr series of f and g are given by

$$F(z_1, z_2, \ldots) = \sum_{n=1}^{\infty} a(n) \exp\left\{-\sum_{k=1}^{q(n)} r_{n,k} z_k\right\}$$

and

$$G(z_1, z_2, \ldots) = \sum_{n=1}^{\infty} b(n) \exp\left\{-\sum_{k=1}^{q(n)} r_{n,k} z_k\right\}.$$

Expressing the $b(n)$ in terms of the $a(n)$ we find

$$G(z_1, z_2, \ldots) = \sum_{n=1}^{\infty} a(n) \exp\left\{-\sum_{k=1}^{q(n)} r_{n,k}(z_k + iy_k)\right\} = F(z_1 + iy_1, z_2 + iy_2, \ldots).$$

Since the real part of $z_n + iy_n$ is the real part of z_n, both series take the same set of values on the lines $x_n = \sigma_0 \beta(n)$. Hence $U_f(\sigma_0) = U_g(\sigma_0)$, as asserted. $\qquad \square$

8.10 The set of values taken by a Dirichlet series in a neighborhood of the line $\sigma = \sigma_0$

Definition. Let $f(s)$ be a general Dirichlet series which converges absolutely for $\sigma > \sigma_a$. Given $\delta > 0$ and σ_0 such that $\sigma_0 - \delta > \sigma_a$, we define the set $W_f(\sigma_0; \delta)$ as follows:

$$W_f(\sigma_0; \delta) = \{f(s): \sigma_0 - \delta < \sigma < \sigma_0 + \delta, -\infty < t < +\infty\}.$$

That is, $W_f(\sigma_0; \delta)$ is the set of values taken by $f(s)$ in the strip

$$\sigma_0 - \delta < \sigma < \sigma_0 + \delta.$$

Also, if $\sigma_0 > \sigma_a$ we define

$$W_f(\sigma_0) = \bigcap_{0 < \delta < \sigma_0 - \sigma_a} W_f(\sigma_0; \delta).$$

Thus, $W_f(\sigma_0)$ is the intersection of the sets of values taken by $f(s)$ in all such strips.

It is clear that $V_f(\sigma_0) \subseteq W_f(\sigma_0)$ since every value taken by $f(s)$ on the line $\sigma = \sigma_0$ is also taken in every strip containing this line. Of course, it may happen that $V_f(\sigma_0) = W_f(\sigma_0)$ or that $V_f(\sigma_0) \neq W_f(\sigma_0)$.

In general, we have:

Theorem 8.14. $V_f(\sigma_0) \subseteq W_f(\sigma_0) \subseteq \overline{V_f(\sigma_0)}$, hence $\overline{V_f(\sigma_0)} = \overline{W_f(\sigma_0)}$.

PROOF. We remark that this proof is entirely function-theoretic and has nothing to do with the concept of a basis.

We are to prove that every point in $W_f(\sigma_0)$ is in the closure of $V_f(\sigma_0)$. We will show that if $w \in W_f(\sigma_0)$ then w is an adherent point of $V_f(\sigma_0)$. In fact, we will prove that

$$w = \lim_{n \to \infty} f(\sigma_0 + it_n)$$

for some real sequence $\{t_n\}$.

Since $w \in W_f(\sigma_0)$ this means that $w \in W_f(\sigma_0; \delta)$ for all $\delta > 0$ such that $\delta < \sigma_0 - \sigma_a$. In particular, $w \in W_f(\sigma_0; 1/n)$ for all $n \geq n_0$ for some n_0. This means that for $n \geq n_0$ we have $w = f(s_n)$ where $s_n = \sigma_n + it_n$ and $\sigma_0 - (1/n) < \sigma_n < \sigma_0 + (1/n)$. Using the numbers t_n so determined, consider the difference

$$w - f(\sigma_0 + it_n) = f(\sigma_n + it_n) - f(\sigma_0 + it_n)$$

where $n \geq n_0$. We shall express this difference in terms of the derivative $f'(s)$. Now just as in the case of ordinary Dirichlet series, the function $f(s)$ defined by

$$f(s) = \sum_{n=1}^{\infty} a(n)e^{-s\lambda(n)}$$

is analytic within its half-plane of absolute convergence. In fact in the proof of Theorem 8.1 we showed that the series converges uniformly on every compact subset of the half-plane $\sigma > \sigma_c$. Therefore the sum is analytic in the half-plane $\sigma > \sigma_c$. Moreover, we can calculate the derivative $f'(s)$ by term-by-term differentiation, so

$$f'(s) = - \sum_{n=1}^{\infty} a(n)\lambda(n)e^{-s\lambda(n)}.$$

Hence if $\sigma \geq \sigma_0$ then s is the half-plane of absolute convergence and we get

$$|f'(s)| \leq \sum_{n=1}^{\infty} |a(n)||\lambda(n)|e^{-\sigma\lambda(n)} = \sum_{n=1}^{\infty} |a(n)|e^{-\sigma_0'\lambda(n)}|\lambda(n)|e^{-(\sigma-\sigma_0')\lambda(n)}$$

where $\sigma_a < \sigma_0' < \sigma_0$. Now $|\lambda(n)|e^{-(\sigma-\sigma_0')\lambda(n)} \to 0$ as $n \to \infty$ so, in particular, this factor is less than 1 for large enough n. Hence

$$|f'(s)| \leq \sum_{n=1}^{\infty} |a(n)|e^{-\sigma_0'\lambda(n)} \cdot K$$

for some K, which shows that $|f'(s)|$ is uniformly bounded in the region $\sigma \geq \sigma_0'$. Let $\sigma_0' = \sigma_0 - 1/n_0$ and let M be an upper bound for $|f'(s)|$ in the region $\sigma \geq \sigma_0'$. Then

$$|w - f(\sigma_0 + it_n)| = |f(\sigma_n + it_n) - f(\sigma_0 + it_n)| = \left| \int_{\sigma_0}^{\sigma_n} f'(\sigma + it_n) \, d\sigma \right|$$

$$\leq M|\sigma_n - \sigma_0| \leq \frac{M}{n}$$

if $n \geq n_0$. Hence $\lim_{n\to\infty} f(\sigma_0 + it_n) = w$, so w is an adherent point of $V_f(\sigma_0)$. This completes the proof. □

8.11 Bohr's equivalence theorem

We have just shown that $W_f(\sigma_0) \subseteq \overline{V_f(\sigma_0)}$. The next theorem shows that this inclusion is actually equality.

Theorem 8.15. *We have*

$$W_f(\sigma_0) = \overline{V_f(\sigma_0)}.$$

The proof of Theorem 8.15 is lengthy and appears in Section 8.12. In this section we show how Theorem 8.15 leads to Bohr's equivalence theorem.

Theorem 8.16 (Bohr's equivalence theorem). *Let f and g be equivalent Dirichlet series with abscissa of absolute convergence σ_a. Then in any open half plane $\sigma > c_1 \geq \sigma_a$ the functions $f(s)$ and $g(s)$ take the same set of values.*

PROOF. Let $S_f(\sigma_1)$ be the set of values taken by $f(s)$ in the half-plane $\sigma > \sigma_1$. Then

$$S_f(\sigma_1) = \bigcup_{\sigma_0 > \sigma_1} V_f(\sigma_0).$$

Now we prove that

$$S_f(\sigma_1) = \bigcup_{\sigma_0 > \sigma_1} W_f(\sigma_0).$$

First of all, we have $S_f(\sigma_1) \subseteq \bigcup_{\sigma_0 > \sigma_1} W_f(\sigma_0)$ because $V_f(\sigma_0) \subseteq W_f(\sigma_0)$. To get inclusion in the other direction, assume $w \in \bigcup_{\sigma_0 > \sigma_1} W_f(\sigma_0)$. Then $w \in W_f(\sigma_0)$ for some $\sigma_0 > \sigma_1$. Hence $w \in W_f(\sigma_0; \delta)$ for all δ satisfying $0 < \delta < \sigma_0 - \sigma_a$. In other words, $f(s)$ takes the value w in every strip $\sigma_0 - \delta < \sigma < \sigma_0 + \delta$ if $0 < \delta < \sigma_0 - \sigma_a$. In particular, when $\delta = \sigma_0 - \sigma_1$, we have $\sigma_0 - \delta = \sigma_1$ so $f(s) = w$ for some s with $\sigma > \sigma_1$. Hence $w \in S_f(\sigma_1)$. This proves that $\bigcup_{\sigma_0 > \sigma_1} W_f(\sigma_0) \subseteq S_f(\sigma_1)$, so the two sets are equal. Therefore, we also have

$$S_g(\sigma_1) = \bigcup_{\sigma_0 > \sigma_1} W_g(\sigma_0).$$

To prove Bohr's theorem it suffices to prove that

$$W_f(\sigma_0) = W_g(\sigma_0)$$

whenever f and g are equivalent. But $f \sim g$ implies

$$U_f(\sigma_0) = U_g(\sigma_0).$$

Hence $\overline{U_f(\sigma_0)} = \overline{U_g(\sigma_0)}$. But, in view of Theorem 8.9, this means

$$\overline{V_f(\sigma_0)} = \overline{V_g(\sigma_0)}.$$

But Theorem 8.15 states that $\overline{V_f(\sigma_0)} = W_f(\sigma_0)$ and $\overline{V_g(\sigma_0)} = W_g(\sigma_0)$ so Bohr's equivalence theorem is a consequence of Theorem 8.15. $\square$

8.12 Proof of Theorem 8.15

To complete the proof of Bohr's equivalence theorem we need to prove Theorem 8.15, which means we must establish the inclusion relation.

(9) $$\overline{V_f(\sigma_0)} \subseteq W_f(\sigma_0).$$

The proof of (9) makes use of two important theorems of analysis which we state as lemmas:

Lemma 1 (Helly selection principle). *Let $\{\theta_{m,n}\}$ be a double sequence of real numbers which is bounded, say*

$$|\theta_{m,n}| < A \quad \text{for all } m, n.$$

Then there exists a subsequence of integers $n_1 < n_2 < \cdots$ with $n_r \to \infty$ as $r \to \infty$, and a sequence $\{\theta_n\}$ of real numbers such that for every $m = 1, 2, \ldots$, we have

$$\lim_{r \to \infty} \theta_{m, n_r} = \theta_m.$$

Note. The important point is that *one* subsequence $\{n_k\}$ works *for every m.* To show the true import of the Lemma, let us see what we can deduce in a trivial fashion. Display the double sequence as an infinite matrix. Consider

179

the first row: $\{\theta_{1,n}\}_{n=1}^{\infty}$. This is a bounded infinite sequence so it has an accumulation point, say θ_1. Hence there is a subsequence $\{n_r\}$ such that $\lim_{r \to \infty} \theta_{1,n_r} = \theta_1$. Similarly, for the second row there is an accumulation point θ_2 and a subsequence n_r' such that $\lim_{n \to \infty} \theta_{2,n_r'} = \theta_2$, and so on. The subsequence $\{n_r'\}$ needed for θ_2 may be quite different from that needed for θ_1. Helly's principle says that *one* subsequence works simultaneously for *all* rows.

PROOF OF LEMMA 1. Let θ_1 be an accumulation point of the first row and suppose the subsequence $\{n_r^{(1)}\}$ has the property that

$$\lim_{r \to \infty} \theta_{1,n_r^{(1)}} = \theta_1.$$

In the second row, consider only those entries $\theta_{2,n_r^{(1)}}$. This is a bounded sequence which has a convergent subsequence with limit θ_2, say. Thus,

$$\lim_{r \to \infty} \theta_{2,n_r^{(2)}} = \theta_2$$

where $\{n_r^{(2)}\}$ is a subsequence of $\{n_r^{(1)}\}$. Repeat the process indefinitely. At the mth step we have a subsequence $\{n_r^{(m)}\}$ which is a subsequence of all earlier subsequences and a number θ_m such that

$$\lim_{r \to \infty} \theta_{m,n_r^{(m)}} = \theta_m.$$

Now define a sequence $\{n_r\}$ by the diagonal process:

$$n_r = n_r^{(r)}.$$

That is, n_1 is the first integer used in the first row, n_2 the second integer used in the second row, etc. Look at the mth row and consider the sequence $\{\theta_{m,n_r}\}$. We assert that

$$\lim_{r \to \infty} \theta_{m,n_r} = \theta_m.$$

Since $n_r = n_r^{(r)}$, after the mth term in this row we have $r > m$ so every integer $n_r^{(r)}$ occurs in the subsequence $n_r^{(m)}$, so from this point on $\{n_r\}$ is a subsequence of $\{n_r^{(m)}\}$ hence $\theta_{m,n_r} \to \theta_m$, as asserted. ☐

Lemma 2 (Rouché's theorem). *Given two functions $f(z)$ and $g(z)$ analytic inside and on a closed circular contour C. Assume*

$$|g(z)| < |f(z)| \quad \text{on } C.$$

Then $f(z)$ and $f(z) + g(z)$ have the same number of zeros inside C.

PROOF OF LEMMA 2. Let $m = \inf\{|f(z)| - |g(z)| : z \in C\}$. Then $m > 0$ because C is compact and the difference $|f(z)| - |g(z)|$ is a continuous function on C. Hence for all real t in the interval $0 \le t \le 1$ we have

$$|f(z) + tg(z)| \ge |f(z)| - |tg(z)| \ge |f(z)| - |g(z)| \ge m > 0.$$

If $0 \leq t \leq 1$ define a number $\varphi(t)$ by the equation

$$\varphi(t) = \frac{1}{2\pi i} \oint_C \frac{f'(z) + tg'(z)}{f(z) + tg(z)} \, dz.$$

This number $\varphi(t)$ is an integer, the number of zeros minus the number of poles of the function $f(z) + tg(z)$ inside C. But there are no poles, so $\varphi(t)$ is the number of zeros of $f(z) + tg(z)$ inside C. But $\varphi(t)$ is a continuous function of t on $[0, 1]$. Since it is an integer, it is constant: $\varphi(0) = \varphi(1)$. But $\varphi(0)$ is the number of zeros of $f(z)$, and $\varphi(1)$ is the number of zeros of $f(z) + g(z)$. This proves Rouché's theorem. $\qquad\square$

PROOF OF RELATION (9). $\overline{V_f(\sigma_0)} \subseteq W_f(\sigma_0)$. Assume $v \in \overline{V_f(\sigma_0)}$. Then either $v \in V_f(\sigma_0)$ or v is an accumulation point of $V_f(\sigma_0)$. If $v \in V_f(\sigma_0)$ then $v \in W_f(\sigma_0)$ since $V_f(\sigma_0) \subseteq W_f(\sigma_0)$. Hence we can assume that v is an accumulation point of $V_f(\sigma_0)$, and $v \notin V_f(\sigma_0)$. This means there is a sequence $\{t_n\}$ of real numbers such that

$$v = \lim_{n \to \infty} f(\sigma_0 + it_n).$$

We wish to prove that $v \in W_f(\sigma_0)$. This means we must show that $v \in W_f(\sigma_0; \delta)$ for every δ satisfying $0 < \delta < \sigma_0 - \sigma_a$. In other words, if $0 < \delta < \sigma_0 - \sigma_a$ we must find an $s = \sigma + it$ in the strip

$$\sigma_0 - \delta < \sigma < \sigma_0 + \delta$$

such that $f(s) = v$. Therefore we are to exhibit an s in this strip such that

$$f(s) = \lim_{n \to \infty} f(\sigma_0 + it_n).$$

Let us examine the numbers $f(\sigma_0 + it_m)$ for the given sequence $\{t_n\}$. We have

$$f(\sigma_0 + it_m) = \sum_{n=1}^{\infty} a(n) e^{-\sigma_0 \lambda(n)} \cdot e^{-it_m \lambda(n)}.$$

The products $t_m \lambda(n)$ form a double sequence. There exists a double sequence of real numbers $\theta_{n,m}$ such that

$$\theta_{n,m} = t_m \lambda(n) + 2\pi k_{m,n}, \text{ with } 0 \leq \theta_{n,m} < 2\pi,$$

where $k_{m,n}$ is an integer. If we replace $t_m \lambda(n)$ by $\theta_{n,m}$ in the series we don't alter the terms, hence

$$f(\sigma_0 + it_m) = \sum_{n=1}^{\infty} a(n) e^{-\sigma_0 \lambda(n)} e^{-i\theta_{n,m}}.$$

By Lemma 1, there is a subsequence of integers $\{n_r\}$ and a sequence of real numbers $\{\theta_m\}$ such that

(10) $$\lim_{r \to \infty} \theta_{m, n_r} = \theta_m.$$

Use this sequence $\{\theta_m\}$ to form a new Dirichlet series

$$g(s) = \sum_{n=1}^{\infty} b(n)e^{-s\lambda(n)}$$

where

$$b(n) = a(n)e^{-i\theta_n}.$$

This has the same abscissa of absolute convergence as $f(s)$. Now consider the following sequence of functions:

$$f_r(s) = f(s + it_{n_r})$$

where $\{n_r\}$ is the subsequence for which (10) holds. We assert that

(a) $f_r(s) \to g(s)$ uniformly in the strip $\sigma_0 - \delta < \sigma < \sigma_0 + \delta$, hence, in particular, in the circular disk $|s - \sigma_0| < \delta$.
(b) $g(\sigma_0) = v$.
(c) There is a $d, 0 < d < \delta$, and an R such that $f_R(s) - v$ and $g(s) - v$ have the same number of zeros in the open disk $|s - \sigma_0| < d$.

If we prove (b) and (c) then $f_R(s) - v$ has at least one zero in the disk because $g(\sigma_0) = v$. But $f_R(s) = f(s + it_{n_R})$ and $s + it_{n_R}$ is in the strip if s is in the disk, so this proves the theorem. Now we prove (a), (b) and (c).

Proof of (a). We have

$$|f_r(s) - g(s)| = \left| \sum_{n=1}^{\infty} a(n)e^{-s\lambda(n)}(e^{-i\theta_{n,n_r}} - e^{-i\theta_n}) \right|$$

$$\leq \sum_{n=1}^{\infty} |a(n)|e^{-\sigma\lambda(n)}|e^{-i\theta_{n,n_r}} - e^{-i\theta_n}|$$

$$\leq \sum_{n=1}^{N} |a(n)|e^{-(\sigma_0-\delta)\lambda(n)}|e^{-i\theta_{n,n_r}} - e^{-i\theta_n}|$$

$$+ 2 \sum_{n=N+1}^{\infty} |a(n)|e^{-(\sigma_0-\delta)\lambda(n)}.$$

Now if $\varepsilon > 0$ is given there is a number $N = N(\varepsilon)$ such that

$$2 \sum_{n=N+1}^{\infty} |a(n)|e^{-(\sigma_0-\delta)\lambda(n)} < \frac{\varepsilon}{2},$$

because the series $\sum_{n=1}^{\infty} |a(n)|e^{-(\sigma_0-\delta)\lambda(n)}$ converges. In the finite sum $\sum_{n=1}^{N}$ we use the inequality

$$|e^{-ib} - e^{-ia}| = \left| \frac{1}{i} \int_a^b e^{-it} \, dt \right| \leq |b - a|$$

to write

$$|e^{-i\theta_{n,n_r}} - e^{-i\theta_n}| \le |\theta_{n,n_r} - \theta_n|.$$

But if $M(\delta) = 1 + \sum_{n=1}^{\infty} |a(n)| e^{-(\sigma_0 - \delta)\lambda(n)}$, there is an integer $r_0 = r_0(\varepsilon)$ such that for every $n = 1, 2, \ldots, N$ we have

$$|\theta_{n,n_r} - \theta_n| < \frac{\varepsilon}{2M(\delta)} \quad \text{if } r \ge r_0.$$

Therefore, if $r \ge r_0$ we have

$$|f_r(s) - g(s)| \le \frac{\varepsilon}{2M(\delta)} \sum_{n=1}^{N} |a(n)| e^{-(\sigma_0 - \delta)\lambda(n)} + \frac{\varepsilon}{2} < \frac{\varepsilon}{2} + \frac{\varepsilon}{2} = \varepsilon.$$

Since r_0 depends only on ε and on δ this shows that $f_r(s) \to g(s)$ uniformly in the strip $\sigma_0 - \delta < \sigma < \sigma_0 + \delta$ as $r \to \infty$. This proves (a).

Proof of (b). We use (a) to write

$$g(\sigma_0) = \lim_{r \to \infty} f_r(\sigma_0) = \lim_{r \to \infty} f(\sigma_0 + it_{n_r}) = v.$$

Proof of (c). Assume first that g is not constant. Since $g(\sigma_0) = v$ there is a positive $d < \delta$ such that $g(s) \neq v$ on the circle

$$C = \{s : |s - \sigma_0| = d\}.$$

Let M be the minimum value of $|g(s) - v|$ on C. Then $M > 0$. Now choose R so large that $|f_R(s) - g(s)| < M$ on C. This is possible by uniform convergence of the sequence $\{f_R(s)\}$, since the circle C lies within the strip $|\sigma - \sigma_0| < \delta$. Then on C we have

$$|f_R(s) - g(s)| < M \le |g(s) - v|.$$

If $G(s) = f_R(s) - g(s)$ and $F(s) = g(s) - v$ we have $|G(s)| < |F(s)|$ on C with $F(s), G(s)$ analytic inside C. Therefore, by Rouché's theorem the functions $F(s) + G(s)$ and $F(s)$ have the same number of zeros inside C. But $F(s) + G(s) = f_R(s) - v$, so $f_R(s) - v$ has the same number of zeros inside C as $g(s) - v$. Now $g(\sigma_0) = v$ so $g(s) - v$ has at least one zero inside C. Hence $f_R(s) - v$ has at least one zero inside C. As noted earlier, this completes the proof if g is not constant.

To complete the proof we must consider the possibility that $g(s)$ is constant in the half-plane of absolute convergence. Then $g'(s) = 0$ for all s in this half-plane, which means

$$g'(s) = -\sum_{n=1}^{\infty} \lambda(n) b(n) e^{-s\lambda(n)} = 0.$$

But as in the case of ordinary Dirichlet series, if a general Dirichlet series has the value 0 for a sequence of values of s with real parts tending to $+\infty$ then all the coefficients must be zero. (See [4], Theorem 11.3.) Hence

183

$\lambda(n)b(n) = 0$ for all $n \geq 1$. Therefore $b(n) = 0$ with at most one exception, say $b(n_1)$, in which case $\lambda(n_1) = 0$. Therefore, since $a(n) = b(n)e^{i\theta_n}$, we must have $a(n) = 0$ with at most one exception, say $a(n_1)$, and then $\lambda(n_1) = 0$. Hence the series for $f(s)$ consists of only one term, $f(s) = a(n_1)e^{-s\lambda(n_1)} = a(n_1)$, so $f(s)$ itself is constant. But in this case the theorem holds trivially. $\qquad\square$

8.13 Examples of equivalent Dirichlet series. Applications of Bohr's theorem to L-series

Theorem 8.17. *Let $k \geq 1$ be a given integer, and let χ be any Dirichlet character modulo k. Let $\sum_{n=1}^{\infty} a(n)n^{-s}$ be any Dirichlet series whose coefficients have the following property:*

$$a(n) \neq 0 \text{ implies } (n, k) = 1.$$

Then

$$\sum_{n=1}^{\infty} \frac{a(n)}{n^s} \sim \sum_{n=1}^{\infty} \frac{a(n)\chi(n)}{n^s}.$$

PROOF. Since these are ordinary Dirichlet series we may use Theorem 8.12 to establish the equivalence. In this case we take $f(n) = \chi(n)$. Then f is completely multiplicative and condition (a) is satisfied. Now we show that condition (b) is satisfied. We need to show that $|f(p)| = 1$ if $a(n) \neq 0$ and $p \mid n$. But $a(n) \neq 0$ implies $(n, k) = 1$. Since $p \mid n$ we must have $(p, k) = 1$ so $|f(p)| = |\chi(p)| = 1$ since χ is a character. Therefore the two series are equivalent. $\qquad\square$

Theorem 8.18. *For a given modulus k, let $\chi_1, \ldots, \chi_{\varphi(k)}$ denote the Dirichlet characters modulo k. Then in any half-plane of the form $\sigma > \sigma_1 \geq 1$ the set of values taken by the Dirichlet L-series $L(s, \chi_i)$ is independent of i.*

PROOF. Applying the previous theorem with $a(n) = \chi_1(n)$ we have

$$\sum_{n=1}^{\infty} \frac{\chi_1(n)}{n^s} \sim \sum_{n=1}^{\infty} \frac{\chi_1(n)\chi(n)}{n^s}$$

for every character χ modulo k. Here we use the fact that $\chi_1(n) \neq 0$ implies $(n, k) = 1$. Thus each L-series $L(s, \chi)$ is equivalent to the particular L-series $L(s, \chi_1)$. Therefore, by Bohr's theorem, $L(s, \chi)$ takes the same set of values as $L(s, \chi_1)$ in any open half-plane within the half-plane of absolute convergence. $\qquad\square$

8.14 Applications of Bohr's theorem to the Riemann zeta function

Our applications to the Riemann zeta function require the following identity involving Liouville's function $\lambda(n)$ which is defined by the relations

$$\lambda(1) = 1, \quad \lambda(p_1^{a_1} \cdots p_r^{a_r}) = (-1)^{a_1 + \cdots + a_r}.$$

The function $\lambda(n)$ is completely multiplicative and we have (see [4], p. 231)

$$\sum_{n=1}^{\infty} \frac{\lambda(n)}{n^s} = \frac{\zeta(2s)}{\zeta(s)} \quad \text{if } \sigma > 1.$$

Theorem 8.19. *Let $\lambda(n)$ denote Liouville's function and let*

$$C(x) = \sum_{n \le x} \frac{\lambda(n)}{n}.$$

Then if $\sigma > 1$ we have

$$\frac{\zeta(2s)}{(s-1)\zeta(s)} = \int_1^{\infty} \frac{C(x)}{x^s} \, dx.$$

PROOF. By Abel's identity (Theorem 4.2 in [4]) we have

$$\sum_{n \le x} \frac{\lambda(n)}{n} \frac{1}{n^s} = \frac{C(x)}{x^s} + s \int_1^x \frac{C(t)}{t^{s+1}} \, dt.$$

Keep $\sigma > 0$ and let $x \to \infty$. Then

$$\frac{C(x)}{x^s} = O\left(\frac{1}{x^{\sigma}} \sum_{n \le x} \frac{1}{n}\right) = O\left(\frac{\log x}{x^{\sigma}}\right) = o(1) \quad \text{as } x \to \infty,$$

so we find

$$\sum_{n=1}^{\infty} \frac{\lambda(n)}{n^{s+1}} = s \int_1^{\infty} \frac{C(t)}{t^{s+1}} \, dt, \quad \text{for } \sigma > 0.$$

Replacing s by $s - 1$ we get

$$\sum_{n=1}^{\infty} \frac{\lambda(n)}{n^s} = (s-1) \int_1^{\infty} \frac{C(t)}{t^s} \, dt \quad \text{for } \sigma > 1.$$

Since the series on the left has sum $\zeta(2s)/\zeta(s)$ the proof is complete. $\qquad \square$

Now we prove a remarkable theorem discovered by P. Turán [50] in 1948 which gives a surprising connection between the Riemann hypothesis and the partial sums of the Riemann zeta function in the half-plane $\sigma > 1$.

Theorem 8.20. *Let*

$$\zeta_n(s) = \sum_{k=1}^{n} \frac{1}{k^s}.$$

If there exists an n_0 such that $\zeta_n(s) \ne 0$ for all $n \ge n_0$ and all $\sigma > 1$, then $\zeta(s) \ne 0$ for $\sigma > \frac{1}{2}$.

PROOF. First we note that the two Dirichlet series $\sum_{k=1}^{n} k^{-s}$ and $\sum_{k=1}^{n} \lambda(k)k^{-s}$ are equivalent because λ is completely multiplicative and has absolute value 1. Therefore, by Bohr's theorem, $\zeta_n(s) \neq 0$ for $\sigma > 1$ implies that $\sum_{k=1}^{n} \lambda(k)k^{-s} \neq 0$ for $\sigma > 1$. But for s real we have

$$\lim_{s \to +\infty} \sum_{k=1}^{n} \frac{\lambda(k)}{k^s} = \lambda(1) = 1.$$

Hence for all real $s > 1$ we must have $\sum_{k=1}^{n} \lambda(k)k^{-s} > 0$. Letting $s \to 1+$ we find

$$\sum_{k=1}^{n} \frac{\lambda(k)}{k} \geq 0 \quad \text{if } n \geq n_0.$$

In other words, the function

$$(11) \qquad\qquad C(x) = \sum_{n \leq x} \frac{\lambda(n)}{n}$$

is nonnegative for $x \geq n_0$. Now we use the identity of Theorem 8.19,

$$\frac{\zeta(2s)}{(s-1)\zeta(s)} = \int_1^{\infty} \frac{C(x)}{x^s}\, dx,$$

valid for $\sigma > 1$. Note that the denominator $(s-1)\zeta(s)$ is nonzero on the real axis $s > \frac{1}{2}$, and $\zeta(2s)$ is finite for real $s > \frac{1}{2}$. Therefore, by the integral analog of Landau's theorem (see Theorem 11.13 in [4]) the function on the left is analytic everywhere in the half-plane $\sigma > \frac{1}{2}$. This implies that $\zeta(s) \neq 0$ for $\sigma > \frac{1}{2}$, and the proof is complete. $\qquad\square$

Turán's theorem assumes that the sum $C(x)$ in (11) is nonnegative for all $x \geq n_0$. In 1958, Haselgrove [14] proved, by an ingenious use of machine computation, that $C(x)$ is negative for infinitely many values of x. Therefore, Theorem 8.20 cannot be used to prove the Riemann hypothesis. Subsequently, Turán [51] sharpened his theorem by replacing the hypothesis $C(x) \geq 0$ by a weaker inequality that cannot be disproved by machine computation.

Theorem 8.21 (Turán). *Let $C(x) = \sum_{n \leq x} \lambda(n)/n$. If there exist constants $\alpha > 0, c > 0$ and n_0 such that*

$$(12) \qquad\qquad C(x) > -c\,\frac{\log^\alpha x}{\sqrt{x}}$$

for all $x \geq n_0$, then the Riemann hypothesis is true.

PROOF. If $\varepsilon > 0$ is given there exists an $n_1 \geq n_0$ such that $c \log^\alpha x \leq x^\varepsilon$ for all $x \geq n_1$ so (12) implies

$$C(x) > -x^{\varepsilon - 1/2}.$$

Let $A(x) = C(x) + x^{\varepsilon - 1/2}$, where ε is fixed and $0 < \varepsilon < \frac{1}{2}$. Then $A(x) > 0$ for all $x \geq n_1$. Also, for $\sigma > 1$ we have

$$\int_1^\infty \frac{A(x)}{x^s} \, dx = \int_1^\infty \frac{C(x)}{x^s} \, dx + \int_1^\infty x^{\varepsilon - s - 1/2} \, dx$$

$$= \frac{\zeta(2s)}{(s-1)\zeta(s)} + \frac{1}{s - \frac{1}{2} - \varepsilon} = f(s),$$

say. Arguing as in the proof of Theorem 8.20, we find that the function $f(s)$ is analytic on the real line $s > \frac{1}{2} + \varepsilon$. By Landau's theorem it follows that $f(s)$ is analytic in the half-plane $\sigma > \frac{1}{2} + \varepsilon$. This implies that $\zeta(s) \neq 0$ for $\sigma > \frac{1}{2} + \varepsilon$, hence $\zeta(s) \neq 0$ for $\sigma > \frac{1}{2}$ since ε can be arbitrarily small. $\qquad\square$

Note. Since each function $\zeta_n(s)$ is a Dirichlet series which does not vanish identically there exists a half-plane $\sigma > 1 + \sigma_n$ in which $\zeta_n(s)$ never vanishes. (See [4], Theorem 11.4.) The exact value of σ_n is not yet known. In his 1948 paper [50] Turán proved that, for all sufficiently large n, $\zeta_n(s) \neq 0$ in the half-plane $\sigma > 1 + 2(\log \log n)/\log n$, hence $\sigma_n \leq 2(\log \log n)/\log n$ for large n. In the other direction, H. L. Montgomery has shown that there exists a constant $c > 0$ such that for all sufficiently large n, $\zeta_n(s)$ has a zero in the half-plane $\sigma > 1 + c(\log \log n)/\log n$, hence $\sigma_n \geq c(\log \log n)/\log n$ for large n.

The number $1 + \sigma_n$ is also equal to the abscissa of convergence of the Dirichlet series for the reciprocal $1/\zeta_n(s)$. If $\sigma > 1 + \sigma_n$ we can write

$$\frac{1}{\zeta_n(s)} = \sum_{k=1}^\infty \frac{\mu_n(k)}{k^s},$$

where $\mu_n(k)$ is the Dirichlet inverse of the function $u_n(k)$ given by

$$u_n(k) = \begin{cases} 1 & \text{if } k \leq n, \\ 0 & \text{if } k > n. \end{cases}$$

The usual Möbius function $\mu(k)$ is the limiting case of $\mu_n(k)$ as $n \to \infty$.

Exercises for Chapter 8

1. If $\sum a(n)e^{-s\lambda(n)}$ has abscissa of convergence $\sigma_c < 0$, prove that

$$\sigma_c = \limsup_{n \to \infty} \frac{\log |\sum_{k=n}^\infty a(k)|}{\lambda(n)}.$$

2. Let σ_c and σ_a denote the abscissae of convergence and absolute convergence of a Dirichlet series. Prove that

$$0 \leq \sigma_a - \sigma_c \leq \limsup_{n \to \infty} \frac{\log n}{\lambda(n)}.$$

This gives $0 \leq \sigma_a - \sigma_c \leq 1$ for ordinary Dirichlet series.

3. If $\log n/\lambda(n) \to 0$ as $n \to \infty$ prove that

$$\sigma_a = \sigma_c = \limsup_{n \to \infty} \frac{\log|a(n)|}{\lambda(n)}.$$

What does this imply about the radius of convergence of a power series?

4. Let $\{\lambda(n)\}$ be a sequence of *complex* numbers. Let A denote the set of all points $s = \sigma + it$ for which the series $\sum a(n)e^{-s\lambda(n)}$ converges absolutely. Prove that A is convex.

Exercises 5, 6, and 7 refer to the series $f(s) = \sum_{n=1}^{\infty} a(n)e^{-s\lambda(n)}$ with exponents and coefficients given as follows

n	1	2	3	4	5
$\lambda(n)$	$-1 - \log 2$	-1	$-\log 2$	$-1 + \log 2$	0
$a(n)$	$\frac{1}{8}$	$\frac{1}{2}$	$\frac{1}{4}$	$-\frac{1}{8}$	$\frac{1}{2}$

n	6	7	8	9	10
$\lambda(n)$	$1 - \log 2$	$\log 2$	1	$\log 3$	$1 + \log 2$
$a(n)$	$\frac{1}{8}$	$-\frac{1}{4}$	$\frac{1}{2}$	$-\frac{3}{4}$	$-\frac{1}{8}$

Also, $a(n + 10) = -\frac{3}{4}2^{-n}$ and $\lambda(n + 10) = (n + 1)\log 3$ for $n \geq 1$.

5. Prove that $\sigma_a = -(\log 2)/\log 3$.

6. Show that the Bohr function corresponding to the basis $B = (1, \log 2, \log 3)$ is

$$F(z_1, z_2, z_3) = \cos(iz_1) - \tfrac{1}{2}i \sin(iz_2)(1 + \cos(iz_1)) + \frac{1 - 2e^{-z_3}}{2 - e^{-z_3}},$$

if $x_3 > -\log 2$, z_1, z_2 arbitrary.

7. Determine the set $U_f(0)$. *Hint:* The points $-1, 1 + i, 1 - i$ are significant.

8. Assume the Dirichlet series $f(s) = \sum_{n=1}^{\infty} a(n)e^{-s\lambda(n)}$ converges absolutely for $\sigma > \sigma_a$. If $\sigma > \sigma_a$ prove that

$$\lim_{T \to +\infty} \frac{1}{2T} \int_{-T}^{T} e^{\lambda(\sigma + it)}f(\sigma + it)\, dt = \begin{cases} a(n) & \text{if } \lambda = \lambda(n) \\ 0 & \text{if } \lambda \neq \lambda(1), \lambda(2), \ldots. \end{cases}$$

9. Assume the series $f(s) = \sum_{n=1}^{\infty} a(n)e^{-s\lambda(n)}$ converges absolutely for $\sigma > \sigma_a > 0$. Let $v(n) = e^{\lambda(n)}$.

(a) Prove that the series $g(s) = \sum_{n=1}^{\infty} a(n)e^{-sv(n)}$ converges absolutely if $\sigma > 0$.
(b) If $\sigma > \sigma_a$ prove that

$$\Gamma(s)f(s) = \int_{0}^{\infty} g(t)t^{s-1}\, dt.$$

This extends the classic formula for the Riemann zeta function,

$$\Gamma(s)\zeta(s) = \int_0^\infty \frac{t^{s-1}}{e^t - 1}\, dt.$$

Hint: First show that $\Gamma(s)e^{-s\lambda(n)} = \int_0^\infty e^{-tv(n)}t^{s-1}\, dt$.

Supplement to Chapter 3

Alternate proof of Dedekind's functional equation

This supplement gives an alternate proof of Dedekind's functional equation as stated in Theorem 3.4:

Theorem. *If* $A = \begin{pmatrix} a & b \\ c & d \end{pmatrix} \in \Gamma$ *and* $c > 0$, *then for every* τ *in* H *we have*

$$(1) \qquad \eta\left(\frac{a\tau + b}{c\tau + d}\right) = \varepsilon(A)\{-i(c\tau + d)\}^{1/2}\eta(\tau),$$

where

$$(2) \qquad \varepsilon(A) = \exp\left\{\pi i\left(\frac{a + d}{12c} - s(d, c)\right)\right\}$$

and $s(d, c)$ *is a Dedekind sum.*

The alternate proof was suggested by Basil Gordon and is based on the fact that the modular group Γ has the two generators $T\tau = \tau + 1$ and $S\tau = -1/\tau$. In Theorem 2.1 we showed that every A in Γ can be expressed in the form

$$A = T^{n_1}ST^{n_2}S \cdots ST^{n_r},$$

where the n_i are integers. But $T = ST^{-1}ST^{-1}S$, so every element of Γ also has the form $ST^{m_1}S \cdots ST^{m_k}$ for some choice of integers $m_1, \ldots, m_k$. The idea of the proof is to show that if the functional equation (1) holds for a particular transformation $A = \begin{pmatrix} a & b \\ c & d \end{pmatrix}$ in Γ with $c > 0$ and with $\varepsilon(A)$ as specified in (2), then it also holds for the products AT^m and AS for every

190

integer m. (See Lemma 3 below.) In Theorem 3.1 we proved that it holds for S. Therefore, because every element of Γ has the form $ST^{m_1}S \cdots ST^{m_k}$, it follows that the functional equation (1) holds for every A with $c > 0$.

The argument is divided into three lemmas that show that the general functional equation is a consequence of the special case in Theorem 3.1 together with three basic properties of Dedekind sums derived in Sections 3.7 and 3.8. The first two lemmas relate $\varepsilon(A)$ with $\varepsilon(AT^m)$ and $\varepsilon(AS)$, where T and S are the generators of the modular group.

Lemma 1. *If* $A = \begin{pmatrix} a & b \\ c & d \end{pmatrix} \in \Gamma$ *and* $c > 0$, *then for every integer* m *we*

have

$$\varepsilon(AT^m) = e^{\pi i m/12}\varepsilon(A).$$

PROOF. We have $AT^m = \begin{pmatrix} a & b \\ c & d \end{pmatrix}\begin{pmatrix} 1 & m \\ 0 & 1 \end{pmatrix} = \begin{pmatrix} a & am + b \\ c & cm + d \end{pmatrix}$, so

$$\varepsilon(AT^m) = \exp\left\{\pi i\left(\frac{a + cm + d}{12c} - s(cm + d, c)\right)\right\}.$$

But $s(cm + d, c) = s(d, c)$ by Theorem 3.5(a), and hence, we obtain Lemma 1. $\qquad\square$

Lemma 2. *If* $A = \begin{pmatrix} a & b \\ c & d \end{pmatrix} \in \Gamma$ *and* $c > 0$, *then we have*

$$\varepsilon(AS) = \begin{cases} e^{-\pi i/4}\varepsilon(A) & \text{if } d > 0, \\ e^{\pi i/4}\varepsilon(A) & \text{if } d < 0. \end{cases}$$

PROOF. We have

$$AS = \begin{pmatrix} a & b \\ c & d \end{pmatrix}\begin{pmatrix} 0 & -1 \\ 1 & 0 \end{pmatrix} = \begin{pmatrix} b & -a \\ d & -c \end{pmatrix}.$$

If $d > 0$, we represent the transformation AS by the matrix

$$AS = \begin{pmatrix} b & -a \\ d & -c \end{pmatrix},$$

but if $d < 0$, then $-d > 0$ and we use the representation $AS = \begin{pmatrix} -b & a \\ -d & c \end{pmatrix}$.

For $d > 0$, we have

(3) $$\varepsilon(AS) = \exp\left\{\pi i\left(\frac{b - c}{12d} - s(-c, d)\right)\right\}$$

$$= \exp\left\{\pi i\left(\frac{b - c}{12d} + s(c, d)\right)\right\}$$

because $s(-c, d) = -s(c, d)$ by Theorem 3.5(a). The reciprocity law for Dedekind sums implies

$$s(c, d) + s(d, c) = \frac{c}{12d} + \frac{d}{12c} - \frac{1}{4} + \frac{1}{12cd}.$$

We replace the numerator 1 in the last fraction by $ad - bc$ and rearrange terms to obtain

$$\frac{b - c}{12d} + s(c, d) = \frac{a + d}{12c} - s(d, c) - \frac{1}{4}.$$

Using this in (3), we find $\varepsilon(AS) = e^{-\pi i/4}\varepsilon(A)$.

If $d < 0$, we use the representation $AS = \begin{pmatrix} -b & a \\ -d & c \end{pmatrix}$ to obtain

(4)
$$\varepsilon(AS) = \exp\left\{ \pi i\left(\frac{-b + c}{-12d} - s(c, -d) \right) \right\}.$$

In this case, $-d' > 0$ and we use the reciprocity law in the form

$$s(c, -d) + s(-d, c) = \frac{c}{-12d} - \frac{d}{12c} - \frac{1}{4} - \frac{ad - bc}{12cd}.$$

Rearrange terms and use $s(-d, c) = -s(d, c)$ to obtain

$$\frac{-b + c}{-12d} - s(c, -d) = \frac{a + d}{12c} - s(d, c) + \frac{1}{4}.$$

Using this in (4), we find that $\varepsilon(AS) = e^{\pi i/4}\varepsilon(A)$. This completes the proof of Lemma 2. $\qquad\qquad\square$

Lemma 3. *If Dedekind's functional equation*

(5)
$$\eta(A\tau) = \varepsilon(A)\{-i(c\tau + d)\}^{1/2}\eta(\tau),$$

is satisfied for some $A = \begin{pmatrix} a & b \\ c & d \end{pmatrix}$ *in* Γ *with* $c > 0$ *and* $\varepsilon(A)$ *given by* (2), *then it is also satisfied for* AT^m *and for* AS. *That is,* (5) *implies*

(6)
$$\eta(AT^m\tau) = \varepsilon(AT^m)\{-i(c\tau + d + mc)\}^{1/2}\eta(\tau),$$

and

(7)
$$\eta(AS\tau) = \varepsilon(AS)\{-i(d\tau - c)\}^{1/2}\eta(\tau) \quad if\ d > 0,$$

whereas

(8)
$$\eta(AS\tau) = \varepsilon(AS)\{-i(-d\tau + c)\}^{1/2}\eta(\tau) \quad if\ d < 0.$$

PROOF. Replace τ by $T'''\tau$ in (5) to obtain

$$\eta(AT'''\tau) = \varepsilon(A)\{-i(cT'''\tau + d)\}^{1/2}\eta(T'''\tau)$$
$$= \varepsilon(A)\{-i(c\tau + mc + d)\}^{1/2}e^{\pi im/12}\eta(\tau).$$

Using Lemma 2 we obtain (6).

Now replace τ by $S\tau$ in (5) to get

$$\eta(AS\tau) = \varepsilon(A)\{-i(cS\tau + d)\}^{1/2}\eta(S\tau).$$

Using Theorem 3.1 we can write this as

(9) $$\eta(AS\tau) = \varepsilon(A)\{-i(cS\tau + d)\}^{1/2}\{-i\tau\}^{1/2}\eta(\tau).$$

If $d > 0$, we write

$$cS\tau + d = -\frac{c}{\tau} + d = \frac{d\tau - c}{\tau};$$

hence,

$$-i(cS\tau + d) = \frac{-i(d\tau - c)}{-i\tau}e^{-\pi i/2},$$

and therefore, $\{-i(cS\tau + d)\}^{1/2}\{-i\tau\}^{1/2} = e^{-\pi i/4}\{-i(d\tau - c)\}^{1/2}$. Using this in (9) together with Lemma 3, we obtain (7).

If $d < 0$, we write

$$cS\tau + d = -\frac{c}{\tau} + d = \frac{-d\tau + c}{-\tau}$$

so that in this case we have

$$-i(cS\tau + d) = \frac{-i(-d\tau + c)}{-i\tau}e^{\pi i/2},$$

and therefore, $\{-i(cS\tau + d)\}^{1/2}\{-i\tau\}^{1/2} = e^{\pi i/4}\{-i(-d\tau + c)\}^{1/2}$. Using this in (9) together with Lemma 3, we obtain (8). $\square$

Remark on the root of unity $\varepsilon(A)$

Dedekind's functional equation (1), with an unspecified 24th root of unity $\varepsilon(A)$, follows immediately by extracting 24th roots in the functional equation for $\Delta(\tau)$. Much of the effort in this theory is directed at showing that the root of unity $\varepsilon(A)$ has the form given in (2). It is of interest to note that a simple argument due to Dedekind gives the following theorem:

Theorem. *If* (1) *holds whenever* $A = \begin{pmatrix} a & b \\ c & d \end{pmatrix} \in \Gamma$ *and* $c \neq 0$, *then*

$$\varepsilon(A) = \exp\left\{\pi i\left(\frac{a + d}{12c} - f(d, c)\right)\right\}$$

for some rational number $f(d, c)$ *depending only on* d *and* c.

PROOF. Let

$$A\tau = \frac{a\tau + b}{c\tau + d} \quad \text{and} \quad A'\tau = \frac{a'\tau + b'}{c\tau + d}$$

be two transformations in Γ having the same denominator $c\tau + d$. Then

$$ad - bc = 1 \quad \text{and} \quad a'd - b'c = 1,$$

so both pairs a, b and a', b' are solutions of the linear Diophantine equation

$$xd - yc = 1.$$

Consequently, there is an integer n such that

$$a' = a + nc, \quad b' = b + nd.$$

Hence,

$$A'\tau = \frac{(a + nc)\tau + (b + nd)}{c\tau + d} = \frac{a\tau + b}{c\tau + d} + n = A\tau + n.$$

Therefore, we have

$$\eta(A'\tau) = \eta(A\tau + n) = e^{\pi i n/12}\eta(A\tau) = e^{\pi i n/12}\varepsilon(A)\{-i(c\tau + d)\}^{1/2}\eta(\tau),$$

because of (1). On the other hand, (1) also gives us

$$\eta(A'\tau) = \varepsilon(A')\{-i(c\tau + d)\}^{1/2}\eta(\tau).$$

Comparing the two expressions for $\eta(A'\tau)$, we find $\varepsilon(A') = e^{\pi i n/12}\varepsilon(A)$. But $n = (a' - a)/c$, so

$$\varepsilon(A') = \exp\left(\frac{\pi i(a' - a)}{12c}\right)\varepsilon(A),$$

or

$$\exp\left(-\frac{\pi i a'}{12c}\right)\varepsilon(A') = \exp\left(-\frac{\pi i a}{12c}\right)\varepsilon(A).$$

This shows that the product $\exp\left(-\dfrac{\pi i a}{12c}\right)\varepsilon(A)$ depends only on c and d. Therefore, the same is true for the product

$$\exp\left(-\frac{\pi i(a + d)}{12c}\right)\varepsilon(A).$$

This complex number has absolute value 1 and can be written as

$$\exp\left(-\frac{\pi i(a + d)}{12c}\right)\varepsilon(A) = \exp(-\pi i f(d, c))$$

for some real number $f(d, c)$ depending only on c and d. Hence,

$$\varepsilon(A) = \exp\left\{\pi i\left(\frac{a + d}{12c} - f(d, c)\right)\right\}.$$

Because $\varepsilon^{24} = 1$, it follows that $12cf(d, c)$ is an integer, so $f(d, c)$ is a rational number. $\qquad\square$

Bibliography

1. Apostol, Tom M. Sets of values taken by Dirichlet's *L*-series. *Proc. Sympos. Pure Math.*, Vol. VIII, 133–137. Amer. Math. Soc., Providence, R.I., 1965. MR 31 #1229.

2. Apostol, Tom M. *Calculus*, Vol. II, 2nd Edition. John Wiley and Sons, Inc. New York, 1969.

3. Apostol, Tom M. *Mathematical Analysis*, 2nd Edition. Addison-Wesley Publishing Co., Reading, Mass., 1974.

4. Apostol, Tom M. *Introduction to Analytic Number Theory*. Undergraduate Texts in Mathematics. Springer-Verlag, New York, 1976.

5. Atkin, A. O. L. and O'Brien, J. N. Some properties of $p(n)$ and $c(n)$ modulo powers of 13. *Trans. Amer. Math. Soc. 126* (1967), 442–459. MR 35 #5390.

6. Bohr, Harald. Zur Theorie der allgemeinen Dirichletschen Reihen. *Math. Ann. 79* (1919), 136–156.

7. Deligne, P. La conjecture de Weil. I. *Inst. haut. Étud sci., Publ. math.* 43 (1973), 273–307 (1974). Z. 287, 14001.

8. Erdös, P. A note on Farey series. *Quart. J. Math.*, Oxford Ser. 14 (1943), 82–85. MR 5, 236b.

9. Ford, Lester R. Fractions. *Amer. Math. Monthly 45* (1938), 586–601.

10. Gantmacher, F. R. *The Theory of Matrices*, Vol. 1. Chelsea Publ. Co., New York, 1959.

11. Gunning, R. C. *Lectures on Modular Forms*. Annals of Mathematics Studies, No. 48. Princeton Univ. Press, Princeton, New Jersey, 1962. MR 24 #A2664.

12. Gupta, Hansraj. An identity. *Res. Bull. Panjab Univ. (N.S.) 15* (1964), 347–349 (1965). MR 32 #4070.

13. Hardy, G. H. and Ramanujan, S. Asymptotic formulae in combinatory analysis. *Proc. London Math. Soc.* (2) *17* (1918), 75–115.

14. Haselgrove, C. B. A disproof of a conjecture of Pólya. *Mathematika 5* (1958), 141–145. MR 21 #3391.

15. Hecke, E. Über die Bestimmung Dirichletscher Reihen durch ihre Funktional-gleichung. *Math. Ann. 112* (1936), 664–699.

16. Hecke, E. Über Modulfunktionen und die Dirichlet Reihen mit Eulerscher Produktentwicklung. I. *Math. Ann. 114* (1937), 1–28; II. 316–351.

17. Iseki, Shô. The transformation formula for the Dedekind modular function and related functional equations. *Duke Math. J. 24* (1957), 653–662. MR 19, 943a.

18. Knopp, Marvin I. *Modular Functions in Analytic Number Theory*. Markham Mathematics Series, Markham Publishing Co., Chicago, 1970. MR 42 #198.

19. Lehmer, D. H. Ramanujan's function $\tau(n)$. *Duke Math. J. 10* (1943), 483–492. MR 5, 35b.

20. Lehmer, D. H. Properties of the coefficients of the modular invariant $J(\tau)$. *Amer. J. Math. 64* (1942), 488–502. MR 3, 272c.

21. Lehmer, D. H. On the Hardy-Ramanujan series for the partition function. *J. London Math. Soc. 12* (1937), 171–176.

22. Lehmer, D. H. On the remainders and convergence of the series for the partition function. *Trans. Amer. Math. Soc. 46* (1939), 362–373. MR 1, 69c.

23. Lehner, Joseph. Divisibility properties of the Fourier coefficients of the modular invariant $j(\tau)$. *Amer. J. Math. 71* (1949), 136–148. MR 10, 357a.

24. Lehner, Joseph. Further congruence properties of the Fourier coefficients of the modular invariant $j(\tau)$. *Amer. J. Math. 71* (1949), 373–386. MR 10, 357b.

25. Lehner, Joseph, and Newman, Morris. Sums involving Farey fractions. *Acta Arith. 15* (1968/69), 181–187. MR 39 #134.

26. Lehner, Joseph. *Lectures on Modular Forms*. National Bureau of Standards, Applied Mathematics Series, 61, Superintendent of Documents, U.S. Government Printing Office, Washington, D.C., 1969. MR 41 #8666.

27. LeVeque, William Judson. *Reviews in Number Theory*, 6 volumes. American Math. Soc., Providence, Rhode Island, 1974.

28. Mordell, Louis J. On Mr. Ramanujan's empirical expansions of modular functions. *Proc. Cambridge Phil. Soc. 19* (1917), 117–124.

29. Neville, Eric H. The structure of Farey series. *Proc. London Math. Soc. 51* (1949), 132–144. MR 10, 681f.

30. Newman, Morris. Congruences for the coefficients of modular forms and for the coefficients of $j(\tau)$. *Proc. Amer. Math. Soc. 9* (1958), 609–612. MR 20 #5184.

31. Petersson, Hans. Über die Entwicklungskoeffizienten der automorphen formen. *Acta Math. 58* (1932), 169–215.

32. Petersson, Hans. Über eine Metrisierung der ganzen Modulformen. *Jber. Deutsche Math. 49* (1939), 49–75.

33. Petersson, H. Konstruktion der sämtlichen Lösungen einer Riemannscher Funktionalgleichung durch Dirichletreihen mit Eulersche Produktenwicklung. I. *Math. Ann. 116* (1939), 401–412. Z. 21, p. 22; II. *117* (1939), 39–64. Z. 22, 129.

34. Rademacher, Hans. Über die Erzeugenden von Kongruenzuntergruppen der Modulgruppe. *Abh. Math. Seminar Hamburg*, 7 (1929), 134–148.

35. Rademacher, Hans. Zur Theorie der Modulfunktionen. *J. Reine Angew. Math. 167* (1932), 312–336.

36. Rademacher, Hans. On the partition function $p(n)$. *Proc. London Math. Soc.* (2) *43* (1937), 241–254.

37. Rademacher, Hans. The Fourier coefficients of the modular invariant $j(\tau)$. *Amer. J. Math. 60* (1938), 501–512.

38. Rademacher, Hans. On the expansion of the partition function in a series. *Ann. of Math.* (2) *44* (1943), 416–422. MR 5, 35a.

39. Rademacher, Hans. *Topics in Analytic Number Theory.* Die Grundlehren der mathematischen Wissenschaften, Bd. 169, Springer-Verlag, New York-Heidelberg-Berlin, 1973. Z. 253.10002.

40. Rademacher, Hans and Grosswald, E. *Dedekind Sums.* Carus Mathematical Monograph, 16. Mathematical Association of America, 1972. Z. 251. 10020.

41. Rademacher, Hans and Whiteman, Albert Leon. Theorems on Dedekind sums. *Amer. J. Math. 63* (1941), 377–407. MR 2, 249f.

42. Rankin, Robert A. *Modular Forms and Functions.* Cambridge University Press, Cambridge, Mass., 1977. MR 58 #16518.

43. Riemann, Bernhard. *Gessamelte Mathematische Werke.* B. G. Teubner, Leipzig, 1892. Erläuterungen zu den Fragmenten XXVIII. Von R. Dedekind, pp. 466–478.

44. Schoeneberg, Bruno. *Elliptic Modular Functions.* Die Grundlehren der mathematischen Wissenschaften in Einzeldarstellungen, Bd. 203, Springer-Verlag, New York-Heidelberg-Berlin, 1974. MR 54 #236.

45. Sczech, R. Ein einfacher Beweis der Transformationsformel für log $\eta(z)$. *Math. Ann. 237* (1978), 161–166. MR 58 #21948.

46. Selberg, Atle. On the estimation of coefficients of modular forms. *Proc. Sympos. Pure Math.*, Vol. VIII, pp. 1–15. Amer. Math. Soc., Providence, R.I., 1965. MR 32 #93.

47. Serre, Jean-Pierre. *A Course in Arithmetic.* Graduate Texts in Mathematics, 7. Springer-Verlag, New York-Heidelberg-Berlin, 1973.

48. Siegel, Carl Ludwig. A simple proof of $\eta(-1/\tau) = \eta(\tau)\sqrt{\tau/i}$. *Mathematika 1* (1954), 4. MR 16, 16b.

49. Titchmarsh, E. C. *Introduction to the Theory of Fourier Integrals.* Oxford, Clarendon Press, 1937.

50. Turán, Paul. On some approximative Dirichlet polynomials in the theory of the zeta-function of Riemann. *Danske Vid. Selsk. Mat.-Fys. Medd. 24* (1948), no. 17, 36 pp. MR 10, 286b.

51. Turán, Paul. Nachtrag zu meiner Abhandlung "On some approximative Dirichlet polynomials in the theory of the zeta-function of Riemann." *Acta Math. Acad. Sci. Hungar. 10* (1959), 277–298. MR 22 #6774.

52. Uspensky, J. V. Asymptotic formulae for numerical functions which occur in the theory of partitions [Russian]. *Bull. Acad. Sci. URSS (6) 14* (1920), 199–218.

53. Watson, G. N. *A Treatise on the Theory of Bessel Functions*, 2nd Edition. Cambridge University Press, Cambridge, 1962.

Index of special symbols

Index of special symbols

Index

Index

Index

Riemann, Georg Friedrich Bernhard, 140, 155, 185, 198
Riemann zeta function, 20, 140, 155, 185, 189
Rouché, Eugène, 180
Rouché's theorem, 180

S

Salié, Hans, 136
Schoeneberg, Bruno, 198
Sczech, R., 61, 198
Selberg, Atle, 136, 198
Serre, Jean-Pierre, 198
Siegel, Carl Ludwig, 48, 198
Simultaneous eigenforms, 130
Spitzenform, 114
Subgroups of the modular groups, 46, 75

T

Tau function, 20, 22, 92, 113, 131
Theta function, 91, 141
Transcendental numbers, 147
Transformation of order n, 122
Transformation formula, of Dedekind, 48, 52
 of Iseki, 54
Turán, Paul, 185, 198
Turán's theorem, 185, 186

U

Univalent modular function, 84
Uspensky, J. V., 94, 198

V

Valence of a modular function, 84
Van Wijngaarden, A., 22
Values, of $J(\tau)$, 39
 of Dirichlet series, 170
Vertices of fundamental region, 34

W

Watson, G. N., 109, 198
Weierstrass, Karl, 6
Weierstrass $\wp$-function, 10
Weight of a modular form, 114
Weight formula for zeros of an entire form, 115
Whiteman, Albert Leon, 62, 198

Z

Zeros, of an elliptic function, 5
Zeta function, Hurwitz, 55
 periodic, 55
 Riemann, 140, 155, 185, 189
Zuckerman, Herbert S., 22

Graduate Texts in Mathematics

(continued from page ii)